DIGITAL ELECTRONICS:

FUNDAMENTALS, APPLICATIONS, AND EXPERIMENTS

Howard M. Berlin

Delaware Technical and Community College
Newark, Delaware

RESTON PUBLISHING COMPANY, INC.
A Prentice-Hall Company
Reston, Virginia

Library of Congress Cataloging in Publication Data

Berlin, Howard M.
    Digital electronics.

    1. Digital electronics.   I. Title.
TK7868.D5B46   1985   621.3815   84–2121
ISBN 0–8359–1311–2

*Editorial/production supervision and interior design
by Norma M. Karlin*

CONTENTS

Chapter 3 CLOCKED LOGIC: LATCHES AND FLIP-FLOPS 67

Chapter 4 MULTIVIBRATORS 103

Chapter 5 SHIFT REGISTERS AND COUNTERS 143

Contents

PREFACE

In today's world, it is virtually impossible not to come into daily contact with some product that uses digital technology. From microwave ovens, digital time clocks and watches, CB radios, and laboratory instruments to video games, all use, at least in part, digital integrated circuits. In fact, without the great strides made in this area over the years, the home personal computer could not be a reality and I would still be writing this book without a word processor.

Because of the widespread importance placed on digital devices today, it is now necessary for students in electronic technology programs to possess a firm understanding of the basics of how digital devices operate and how they are combined to create entire systems.

This book is different from most in that it integrates a variety of laboratory experiments along with the basic text material in a single bound volume. In nine chapters, all the basic elements of digital logic are discussed: gates, flip-flops, multivibrators, shift registers, counters, code converters, multiplexers, demultiplexers, digital displays, and memories. As the majority of digital circuits use either TTL or CMOS devices, the interfacing requirements necessary to connect CMOS and TTL devices together are also presented. The tenth chapter discusses several specialized devices that can replace as much as 50 or so discrete logic devices by usually one or two comparatively low cost packages. In most cases it is not necessary to reinvent the wheel, but rather to be aware of what devices, when available, can do the job more economically and efficiently.

vii

Three appendixes are included as useful reference aids for the reader. Appendix A lists those TTL and CMOS devices that are functionally equivalent, assisting the user to easily convert circuits from one logic family to the other. Appendix B gives the pin diagrams of 124 of the most commonly used TTL and CMOS devices. Finally, Appendix C lists the parts and equipment needed to perform all the experiments presented in this book.

Traditionalists may take issue with the fact that a discussion of sacred topics like Boolean algebra, Venn and Veitch diagrams, and optimization techniques such as Karnaugh mapping have been *intentionally* omitted from this book. However, most gate packages cost on the order of 20 to 30 cents apiece and generally contain four individual gates, so optimization techniques are often not worth the effort. In addition, specific exercises at the end of each chapter are not included. We find that when a popular text is repeatedly used over a period of time, answers to these questions are readily available from students who have previously taken the course, even at a two-year institution. For this reason, the course instructor is free to develop and vary exercises to suit the needs of the course, just as examination questions are periodically changed.

Generally, most graduates of two-year technology programs are *troubleshooting* digital circuits rather than actively *designing* them. On the other hand, when those technology graduates have the occasion to design digital circuits, they are generally "one-of-a-kind" circuits or those whose complexity does not warrant the added effort of mapping, optimization, and so on. Consequently, a major emphasis in this book has been placed on discussing the operation of a variety of TTL and CMOS digital integrated-circuit devices designed to accomplish a specific task rather than presenting drawn out explanations of all the ways discrete gates and flip-flops are used to build these same higher-order logic elements, such as digital counters. Since we cannot physically "look" inside digital integrated-circuit packages, it is perhaps more constructive to discuss how the output of a given device responds to various input signals.

With the exception of the first introductory chapter, a series of step-by-step experiments designed to reinforce some of the important concepts discussed are given at the end of each chapter. Depending on the student, several experiments can be completed in a two- or three-hour laboratory session. Although there are a total of 27 experiments, this number was intentionally kept to a minimum for a variety of reasons. At Delaware Technical and Community College, the students in either our electronics, process instrumentation, or robotics programs are responsible for providing their own electronics parts, such as a VOM/DVM, transistors, digital ICs, and resistors, for all laboratory courses in these programs, although other training devices and laboratory instruments are provided. For this reason, the number of components, as well as the total cost of these parts, is kept as low as possible while still affording the students a practical and meaningful learning experience.

This combination text and laboratory workbook is designed to be used by a wide audience. For the electronics student in a two-year technology program, this book can be used in either a one- or two-term introductory course in digital electronics, generally taken during the student's second year after completing course work in basic electricity and analog electronics, but prior to studying microcomputers. This book can also be used by the increasing number of four-year college and university courses in digital electronics, which traditionally are the domain of the electrical engineering departments, but are also now taught by faculty members in other diverse areas such as chemistry and physics where a high degree of theoretical design expertise is not required or desired of their students in such an introductory course. Finally, this book is also well suited for those who like to learn by self-study without requiring an extensive amount of prerequisite background in electricity or electronics other than a knowledge of Ohm's law coupled with the motivation to learn by performing the experiments.

Howard M. Berlin
Wilmington, Delaware

GETTING STARTED

INTRODUCTION

This chapter provides a brief introduction to digital electronics. Before discussing the various types of digital logic elements, we must first consider some ground rules for integrated circuits, logic levels, and binary numbers and what is expected from the experiments found at the end of the following nine chapters.

OBJECTIVES

After completing this chapter you will be able to:

- *Identify logic 0 and logic 1 levels.*
- *Identify the leading and trailing edges of a digital pulse.*
- *Identify the numbered pins on an integrated-circuit device.*
- *List some of the major precautions that must be followed when working with TTL and CMOS devices.*
- *Convert a binary number to its equivalent decimal number.*
- *Understand 4-bit BCD and hexadecimal code systems.*

1

LOGIC SIGNALS

Unlike analog electronics, power supply voltages, as well as signals to and from digital devices, are basically concerned with only two voltage levels. Unless stated otherwise, we will only consider voltage levels that are either at ground (i.e., 0 volts) or at the positive supply voltage that is used to power the digital devices. This convention is termed *positive logic.* When a digital signal is at ground at any point in time, it is said to be at or equal to *logic* 0, or simply 0. On the other hand, if the signal level equals the positive supply voltage, it is said to be at or equal to *logic* 1 or 1. Other terms have been used to indicate these 0 and logic 1 levels. They are:

Logic 0:   Low, False, or Off
Logic 1:   High, True, or On

Although a voltage level of 0 volts (ground) may be referred to as either 0, logic 0, low, false, or off, the practice is to use the correct corresponding term when referring to the logic level that corresponds to the positive supply voltage. For example, if you use the term logic 0, then you should also use the term *logic 1* instead of on, high, or true. Consequently, you should describe logic signals in pairs as either 0 and 1, logic 0 and 1, low and high, false and true, or off and on. In the experiments that are included at the end of the chapters, very often an input is designated either as 0 or 1 to indicate that this input should be connected either to logic 0 (ground) or to logic 1 (the positive supply voltage).

Figure 1–1a is the representation of a single digital signal that goes from a logic 0 to a logic 1 and back to a logic 0 state. Such a signal is called a *positive pulse* and is normally at logic 0. On the other hand, a *negative pulse,* shown in Figure 1–1b, is generated when its level goes from a normally logic 1 state to logic 0 and back to logic 1.

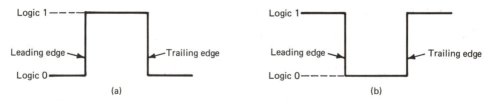

Fig. 1-1 Pulse wave forms showing leading and trailing edges: (a) positive pulse, (b) negative pulse.

Each pulse is characterized by two edges: a *leading* edge and a *trailing* edge. When referring to positive pulses, the leading edge is the logic 0 to 1 transition, whereas the following logic 1 to 0 transition is the trailing edge. For negative pulses, the situation is reversed.

ABOUT INTEGRATED CIRCUITS

Logic Families

Digital integrated-circuit devices are manufactured by different methods and are divided into classes called *logic families*. Digital devices that belong to a given logic family generally have similar electrical characteristics such as power supply levels and speed, and are usually different from devices that belong to other logic families. In this section, the TTL and CMOS logic families are briefly discussed. Additional TTL and CMOS family characteristics are discussed in Chapter 8.

1. TTL Devices

Perhaps the most widely used logic family is *transistor-transistor logic*, or TTL, whose basic internal element is the bipolar transistor. As digital technology has evolved, TTL devices have been further divided into TTL subfamilies that exhibit trade-offs between power consumption and speed. However, all TTL logic families have one thing in common; they all operate with a +5-volt (V) power supply. Consequently, logic 0 is ground, while logic 1 is equivalent to a voltage level of +5 V. Generally, the 5-V level may vary by as much as 5%, so TTL devices may operate properly from any supply voltage from +4.75 to +5.25 V.

There are five widely used TTL subfamilies:

- Standard TTL
- High-power TTL
- Low-power TTL
- Schottky TTL
- Low-power Schottky TTL

Of these five, the standard and low-power Schottky TTL devices are the most popular and are discussed in the following sections.

• Standard TTL

Standard TTL, also called regular TTL or simply TTL, is the oldest, lowest-priced, and most frequently used logic family. Shown in Figure 1–2 is a circuit for a TTL inverter. For the purposes of this book, it is not important to understand how this circuit works, but to realize what is needed to build even the simplest of logic elements: four NPN transistors, two diodes, and four resistors. From this basic arrangement, other, more complicated logic functions are made.

• Low-Power Schottky TTL

The low-power Schottky TTL family of devices, frequently referred to as LS TTL, as its name implies, is intended for low-power consumption. The equivalent circuit for an LS TTL inverter is the same as shown in Figure 1–2, except that Schottky diodes are used in place of the regular diodes, and the four resistors have higher values than the standard TTL circuit, which in turn reduces the current and therefore the power consumption. However, this reduction in power has a penalty. The time that it takes for logic levels to change, the *switching time*, is increased. Therefore, the advantage gained over standard TTL of lower power consumption entails lower speed.

• Decoupling Capacitors

As a rule, all TTL devices have the ability to generate large current spikes on the power supply line when their outputs are switching from

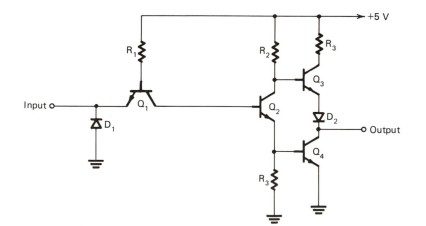

Fig. 1-2 Circuit for a TTL inverter.

byp ass

one state to the other. To remove these unwanted spikes, which may act as digital signals and can possibly upset various stages, *coupling*, or *despiking*, capacitors are placed as close as possible to the integrated-circuit chip and are connected with the shortest leads possible from the device's positive supply line and ground. These capacitors range typically from 0.01 to 0.1 microfarad (μF).

In general, one decoupling capacitor is used for every three or four logic gate packages; otherwise, one capacitor for every two DIP devices other than gates is used. If noise problems persist, more are added.

2. CMOS Devices

Instead of using a bipolar transistor, *complementary metal oxide semiconductor* (CMOS) devices use both p- and n-channel enhancement MOS field-effect transistors (i.e., MOSFETs) as the basic element, as illustrated by the CMOS inverter circuit of Figure 1–3. As a result, this logic family exhibits very low power consumption but is slower than standard TTL devices. However, unlike any of the TTL logic families, the supply and logic levels are not restricted to a fixed voltage, such as 5 V. In fact, supply voltages for CMOS devices can typically range from 3 to 18 V, and, coupled with their low power requirements, make them ideally suited for circuits designed for portable operation using low-cost batteries.

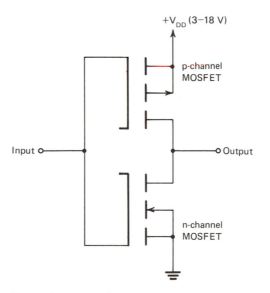

Fig. 1–3 Circuit for a CMOS inverter.

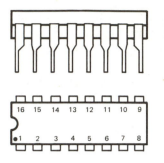

Fig. 1-4 Side and top views of an integrated circuit dual in-line package.

Package Style

Virtually all the digital integrated-circuit devices that you will encounter are packaged in what is termed a *dual in-line package*, or DIP standard. These devices, or chips, as shown in Figure 1–4, generally look like little black or gray "bugs" or centipedes, and, depending on the device, are available with 8, 14, 16, 18, 24, 28, or 40 pins. For example, a 14-pin device has 7 equally spaced pins on each side, and each pin is numbered from 1 to 14 in a *counterclockwise* direction, starting with the pin in the upper-left corner of the chip.

To correctly determine which is the "upper-left corner," each DIP has either a dotlike indentation in one corner or a semicircular cutout at one end, or both. Consequently, one end is somehow different from the other end, which is plain. The end that has the dot or cutout we will call the "top." The pin to the left of the dot or cutout is pin 1. It is very important to be able to identify which is pin 1. From it you must count counterclockwise to find other specific pin numbers. You are almost certain to destroy a device if you connect the power connections backward and apply power or other signals to the wrong pins.

Marked on the top of each device are generally two numbers. One is the device number, while the other number is the date on which the particular device was manufactured. *Standard* TTL devices generally have their serial numbers in the form 74××× and may have a two-letter prefix, such as DM, MM or SN. As an example, the same TTL device may be marked either as 7492, SN7492, or DM7492. Other letters are used to indicate the TTL logic family. For example, a low-power Schottky TTL version of the 7492 device may be marked as 74LS92, SN74LS92, or DM74LS92. In addition to the basic device number, there may also be a one-letter suffix, usually the letter N, which denotes that the package case is plastic.

The 74 series of TTL devices is generally referred to as the *commercial* version. High-reliability versions designed for use in military systems can operate over a wider temperature range and are marked with a 54 prefix. As an example, the 5492 is the high-reliability counterpart of the 7492 device and is exactly pin for pin the same. CMOS devices may be numbered using either a 4000, 34000, or 74C series code. The 4000 series devices may have a prefix such as CD or MM.

It is very important to identify the device number and not confuse it with the date of manufacture, which is simply a four-digit number that indicates the year and week. For example, as shown in Figure 1–5, a digital integrated-circuit device may have the following markings:

SN7432N

7419

This tells us that the device is a 7432 in a plastic package, and that it was made in 1974 during the 19th week. In most cases the device number is easy to determine. Also, as a general rule, the device number is placed *above* the date of manufacture, and pin 1 is located at the lower-left corner from these numbers.

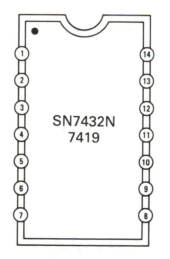

Fig. 1-5 Location of integrated-circuit digital device and date codes.

Some Precautions

Both TTL and CMOS devices can be fun to work with if you keep several basic rules in mind. First, *all* inputs must go somewhere. For TTL devices an unconnected input assumes the equivalent of a logic 1 state. For CMOS devices, inputs of unused gate sections should be connected either to ground or the positive supply voltage. Failure to do so increases the power consumption of the device and eventually may destroy it.

The logic-level transitions for CMOS devices should be fast, typically less than 5 microseconds (μs). Because of their extremely high input impedance and thus sensitive to static discharges, CMOS devices should be handled with caution, preferably while grounded. When not in use, CMOS devices should be stored in conductive foam or on a metal tray. In addition, CMOS devices should not have input signals applied if the power is disconnected. Conversely, CMOS devices should not have their inputs disconnected when power is applied unless a pull-up resistor is connected to the input.

BINARY NUMBERS AND CODES

In the world of digital logic, the simplest *code*, or way to represent data, is the binary code. The binary code is a *two-state* code, since a given signal can only be either at logic 0 or logic 1 at any given time. Using the binary code, the decimal number 0 is represented by a logic 0, while the decimal number 1 is represented by a logic 1.

Using binary numbers, functional elements such as decoders, encoders, and counters primarily rely on a coding system representing the decimal numbers 0 through 9. Such a code is called *8–4–2–1 binary weighted code*, which is frequently referred to as *binary coded decimal*, or *BCD*. Each of the ten numbers from 0 to 9 is represented, or *encoded*, by *four* binary numbers, or *4-bit* binary code, as summarized in Table 1–1.

An extension of the BCD system allows for the representation of 16 numbers, 0 through 15, also by 4-bit binary numbers, called the *hexadecimal* system (Table 1–2). Throughout this book, the BCD code will be primarily used, although some instances will require hexadecimal binary numbers.

In many circumstances, a binary number can be represented by as many bits as necessary, such as the 8-bit number 10011101. To translate any binary number to an equivalent *decimal* number, we

Table 1-1
Binary Coded Decimal Code

Decimal Number	4-Bit BCD Code
0	0000
1	0001
2	0010
3	0011
4	0100
5	0101
6	0110
7	0111
8	1000
9	1001

Table 1-2
Hexadecimal Code

Decimal Number	4-Bit Hexadecimal Code
0	0000
1	0001
2	0010
3	0011
4	0100
5	0101
6	0110
7	0111
8	1000
9	1001
10	1010
11	1011
12	1100
13	1101
14	1110
15	1111

must use powers of 2. As an example, the 4-bit binary code 1010 can be broken down into powers of 2 so that

$$1010_2 = (1 \times 2^3) + (0 \times 2^2) + (1 \times 2^1) + (0 \times 2^0)$$
$$= 8 + 0 + 2 + 0$$
$$= 10$$

Therefore, the binary number 1010 equals the decimal number 10. We start with the rightmost binary number, which in this case is 0, and multiply it by 2 raised to the *zero* power. Then, moving to the left, the next binary number (i.e., 1) is multiplied by 2 raised to the power that is one more than the previous number, which in this case is 2^1, and so on. To speed up this series of calculations, powers that are multiplied by binary 0s are normally omitted, as the result is zero. Consequently, the binary number 1010 can be calculated as

$$1010_2 = 2^3 + 2^1$$
$$= 8 + 2$$
$$= 10$$

Now using the 8-bit number 10011101, its equivalent decimal number can be calculated:

$$10011101_2 = 2^7 + 2^4 + 2^3 + 2^2 + 2^0$$
$$= 128 + 16 + 8 + 4 + 1$$
$$= 157$$

In general, for an *N*-bit binary code, the maximum equivalent decimal number is $2^N - 1$. Consequently, for an 8-bit binary code, the maximum decimal number is $2^8 - 1$, or 255. Therefore, an 8-bit code can be used to represent any decimal number from 0 to 255.

RULES FOR SETTING UP THE EXPERIMENTS

Throughout this book, you will have the opportunity to breadboard a variety of circuits, either by using commercial breadboarding aids or by constructing some of the necessary hardware. Before you set up any experiment, it is recommended that you do the following:

- Plan your experiment beforehand. Know what types of results you are expected to observe.

- Disconnect or turn off *all* power and external signal sources from the breadboard.
- Clear the breadboard of all wires and components from previous experiments, unless instructed otherwise.
- Check the wired-up circuit against the schematic diagram to make sure that it is correct.
- When finished, make sure that you disconnect everything *before* you clear the breadboard of wires and components.

FORMAT FOR THE EXPERIMENTS

The instructions for each experiment are presented in the following format:

• Purpose

The material under this heading states the purpose for performing the experiment. It is well for you to have this intended purpose in mind as you conduct the experiment.

• Pin Configuration of Integrated-Circuit Devices

The pin configurations are given under this heading for all integrated-circuit devices used in the experiment.

• Required Parts and Equipment

A listing of the required circuit components and test equipment necessary for the experiment is given under this heading. A list of the components that are needed for all the experiments is given in Appendix C.

• Schematic Diagram of Circuit

Under this heading is the schematic diagram of the completed circuit that you will wire up in the experiment. You should analyze this diagram in an effort to obtain an understanding of the circuit *before* you proceed further.

• **Design Basics**

Under this heading is a summary of the design equations, when applicable, that apply to the design and operation of the circuit.

• **Steps**

A series of sequential steps describes the detailed instructions for performing portions of the experiment. Questions are also included at appropriate points. Any numerical calculations are performed easily on many of the pocket-type calculators.

BREADBOARDING

The breadboard is designed to accommodate the experiments that you will perform in the chapters to follow. The various integrated-circuit devices, resistors, capacitors, and other components, as well as power and signal connections, all tie directly to the breadboard. Figure 1–6 shows the top view of a solderless breadboarding socket, which is manufactured by several companies. When breadboarding, try to keep the following rules in mind:

• Only No. 22, 24, or 26 insulated wire should be used, and it *must* be solid, not stranded.
• Never insert too large a wire or component lead into a breadboarding terminal.

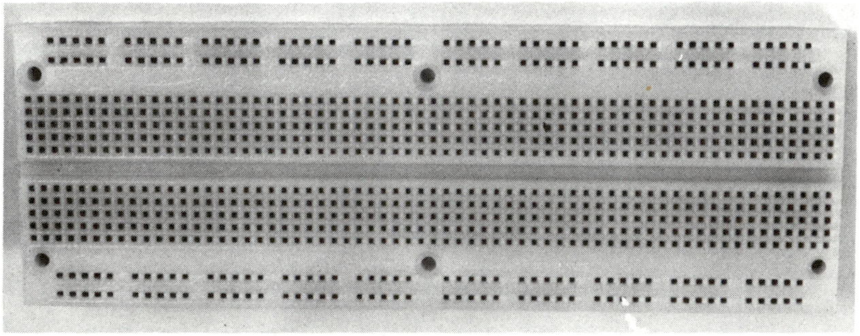

Fig. 1-6 Top view of a solderless bread-boarding socket, which is made by several companies.

- Never insert a bent wire. Straighten out the bent end with a pair of pliers before insertion.
- Try to maintain an orderly arrangement of components and wires, keeping all connections as short as possible.

HELPFUL HINTS AND SUGGESTIONS

Only three tools are really necessary for all the experiments given in this book:

1. A pair of long-nosed pliers.
2. A wire stripper-cutter.
3. A small screwdriver.

The pliers are used to:

- Straighten out the bent ends of hookup wire that is used to wire the circuits on the breadboard.
- Straighten out or bend the resistor, capacitor, and other component leads to the proper position so that they can be conveniently inserted into the breadboard.

The wire stripper–cutter is used to cut the hookup wire to size and strip about 3/8 inch of insulation from each end.

The screwdriver, if for nothing else, can be used to easily remove the integrated-circuit device from the solderless breadboarding socket by gently prying it loose.

REQUIRED EQUIPMENT

Several pieces of equipment will be required for the experiments.

- **Oscilloscope**

Just about any general-purpose type of oscilloscope will do, and it should be a dual-trace type.

• VOM, VTVM, or DVM

A general-purpose type capable of measuring dc voltages is necessary. If you can obtain one, use a digital type; otherwise, any VOM used should have at least a 20 kilohm per volt (kΩ/V) rating so as not to introduce serious loading errors.

• Frequency Counter

This does not have to be an expensive one, but it should have a resolution of 1 hertz (Hz) for precise measurements. Several low-cost units are available in kit form for less than $80.

• Breadboarding Equipment

To perform the experiments, it will be necessary to have several useful circuits for providing input signals and the display of output signals. A number of commercial digital "trainers" are available, such as those manufactured by Heathkit (Figure 1-7) and E&L Instruments, which include most of these functions. Since the experiments will use TTL and CMOS devices, these trainers must be compatible with TTL as well as CMOS devices.

Fig. 1-7 Heathkit ET-3200A digital trainer.

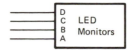

Fig. 1-8 Schematic symbol for a series of LED monitors

1. LED Monitors

A *light-emitting diode (LED) monitor* is a visual monitor, where an LED is lit for a logic 1 state and unlit for a logic 0 state. It is suggested that you have at least four LED monitors. When two or more LED monitors are used, the schematic symbol of Figure 1–8 will be shown in the *Schematic Diagram of Circuit* section of the experiment.

2. Logic Switches

A *logic switch* is a mechanical switch that applies either a logic 0 or 1 at its output terminal. You will usually need logic switches in groups of four, primarily representing the inputs of a 4-bit BCD number. Figure 1–9 shows the schematic diagram that will be used to represent a series of logic switches.

3. Debounced Switches (Pulsers)

In a *mechanical* logic switch, contact bounce (i.e., the uncontrolled making and breaking of contact when the switch contacts are opened or closed) is a common occurrence. In most digital applications, it is extremely important that the output from a switch be bounce free, or *debounced.* A debounced switch or pulser has complementary outputs and will be represented by the schematic symbol of Figure 1–10 when used.

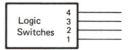

Fig. 1-9 Schematic symbol for a series of logic switches.

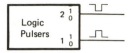

Fig. 1-10 Schematic symbol for a pair of logic pulsers with complementary outputs.

4. Square-Wave Clock

A *square-wave clock* is an astable multivibrator that produces a signal that alternates between logic 0 and 1 at a given rate (frequency). This output frequency may be fixed, using external capacitors, or variable using a potentiometer. When required, the schematic symbol of Figure 1–11 will be used.

5. Seven-Segment LED Display

To easily determine the state of binary information, we can use a *seven-segment LED display* instead of interpreting the states of four separate LED monitors. When needed, the schematic symbol of Figure 1–12 will be used.

6. Function Generator

A function generator is a signal source capable of producing sine waves of variable frequency and amplitude. When needed, the schematic symbol of Figure 1–13 will be used.

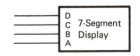

Fig. 1-11 Schematic symbol for a square-wave clock, or astable multivibrator.

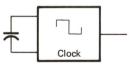

Fig. 1-12 Schematic symbol for a seven-segment LED display.

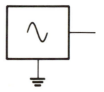

Fig. 1-13 Schematic symbol for a function generator with a sine-wave output.

REVIEW

This chapter presented a brief introduction to the world of digital electronics; some basic ground rules about integrated circuits and binary numbers were discussed. This chapter also discussed what is required for the experiments found at the end of the remaining nine chapters. With this in mind, we are now ready to begin our study of digital electronics in the next chapter by considering the basic building block of digital logic, the gate.

2

DIGITAL GATES AND TRUTH TABLES

OBJECTIVES

After completing this chapter, you will be able to understand:

- *The operation of the inverter.*
- *The operation of AND, NAND, OR, and NOR, XOR, and XNOR gates.*
- *How to enable and disable gate signals.*
- *Gate propagation delay.*
- *The operation of transmission gates.*
- *How to form adders, subtractors, and comparators using basic gate elements.*

GATES AND TRUTH TABLES

Perhaps the simplest of all digital logic building blocks are the gate functions. Simply stated, a gate has two or more inputs and a single output. As we shall see in this section, there are six basic types of logic gates: AND, NAND, OR, NOR, exclusive OR, and exclusive NOR. From

19

these basic gate functions one can build virtually any type of advanced digital logic function. One basic function, which is not thought of as a gate in the traditional sense, is the inverter. Since it is frequently used with gates, its operation will also be discussed in this chapter.

Inverter

The inverter, whose logic symbol is shown in Figure 2–1, can be thought of as a 1-input gate. The output of an inverter is the opposite of its input, so the inverter, as its name implies, *inverts* the input signal. Since only two input signals are possible, logic 0 or 1, the possible output states of an inverter can be summarized as follows:

Input A	Output Q
0	1
1	0

By showing the relation of the output level obtained with the various input combinations as in the preceding tabular form, we have obtained a *truth table*. When a logic device has a number of simultaneous inputs, truth tables can then be quite helpful in describing how a particular device works.

Mathematically, we can express the operation of the inverter as

$$\text{Output} = \text{NOT (input)}$$

where the NOT operator means that we take the *complement* or the opposite of the input. On the schematic symbol shown in Figure 2–1, the circle to the right of the triangle, called an *inversion circle*, tells us that the complement, or NOT operation, is to be performed. If the inversion circle were not present, we would then have a *buffer*, where the output has the same logic state as the input. At first this might not sound too exciting, but buffers serve useful purposes and are discussed in

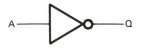

Fig. 2-1 Inverter logic symbol.

Table 2-1
TTL and CMOS Inverters

Inverters per Package	TTL Series	CMOS Series
6	7004	4009, 4049, 74C04

Chapters 7 and 8. Table 2–1 summarizes the TTL and CMOS inverters that are available.

AND Gate

The simplest AND gate that is possible is the 2-input AND gate, whose logic symbol is shown in Figure 2–2. Unlike the inverter, we now have provisions for two separate inputs, *A* and *B*, and a single output signal *Q*. Since we are now dealing with two independent inputs, each of which can have two possible logic states, we can have four possible input combinations:

Input	
A	*B*
0	0
1	0
0	1
1	1

In fact, for logic gates having *N* inputs, each of which has only two possible states (logic 0 or 1), the total number of possible input combinations is 2^N. Consequently, for a 2-input AND gate, we have 2^2, or 4 possible input combinations.

The operation of an AND gate is such that *the output is 1 only when both inputs of a 2-input AND gate are 1.* In other words:

If inputs *A* AND *B* are both 1, then the output will be 1.

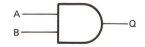

Fig. 2-2 Two-input AND gate logic symbol.

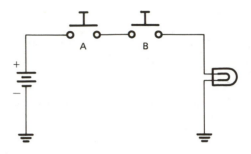

Fig. 2-3 Equivalent switch circuit of a 2-input AND gate.

Consequently, if either input is 0, then the output will be 0.

Another way of explaining the operation of a 2-input AND gate is to visualize the gate as a series of two normally open push-button switches that connect a battery to a light bulb, as shown in Figure 2–3. When both switches are released, that is, have open contacts (0 or OFF), the light is OFF or 0. Only when both switches are simultaneously pressed (1 or ON) is the light ON or 1. We can therefore summarize the operation of a 2-input AND gate by the truth table shown in Table 2–2.

The operation of the preceding 2-input AND gate can now be expanded to include additional inputs, such as the 4-input AND gate with its associated truth table shown in Figure 2–4. Since there are four inputs, each of which can either be 0 or 1, we have 16 possible input combinations. From the description of the operation of the 2-input AND gate, we can now state that the output of *any* AND gate will be 1 only if all inputs are 1. Therefore, the AND gate has only one unique output state, that is, a logic 1 output only when all inputs are 1.

Table 2-2
Two-Input AND
Gate Truth Table

Input		Output
A	*B*	*Q*
0	0	0
1	0	0
0	1	0
1	1	1

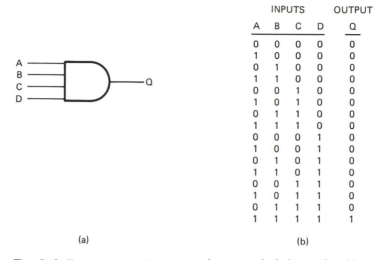

INPUTS				OUTPUT
A	B	C	D	Q
0	0	0	0	0
1	0	0	0	0
0	1	0	0	0
1	1	0	0	0
0	0	1	0	0
1	0	1	0	0
0	1	1	0	0
1	1	1	0	0
0	0	0	1	0
1	0	0	1	0
0	1	0	1	0
1	1	0	1	0
0	0	1	1	0
1	0	1	1	0
0	1	1	1	0
1	1	1	1	1

(a)   (b)

Fig. 2-4 Four-input AND gate: (a) logic symbol, (b) truth table.

Using TTL and CMOS logic families, Table 2–3 lists the various AND gate devices that are available. As an example, the 7411, which is a TTL 3-input AND gate, has three such gates in a single package, as shown in the pin diagram of Figure 2–5. The 4073 is the CMOS *functional* equivalent of the 7411, but it is not necessarily a *pin-for-pin* equivalent, such as is the case with the 74C11.

NAND Gate

The operation of the NAND gate is the opposite of the AND gate. The acronym NAND stands for NOT-AND, meaning that its operation is the NOT, or complement (i.e., opposite), of the AND gate, and this added NOT feature is represented by the inversion circle added to the AND symbol, as shown in Figure 2–6a. Therefore, for a 2-input NAND gate,

Table 2-3
TTL and CMOS AND Gates

Inputs	Gates per Package	TTL Series	CMOS Series
2	4	7408	4081, 74C08
3	3	7411	4073, 74C11
4	2	7421	4082

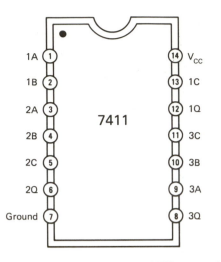

Fig. 2-5 7411 TTL 3-input AND gate pin diagram.

the resulting truth table of Figure 2–6b shows that the output for the NAND gate is the opposite of the AND gate. Consequently, if both inputs *A* and *B* are 1, then the output of the 2-input NAND gate is "NOT the output of the AND gate," which is then 0.

Another way of explaining the operation of a 2-input NAND gate is to visualize the gate as a parallel combination of two normally closed push-button switches that connect a battery to a light bulb, as shown in Figure 2–7. When both switches are released, or contacts are closed (0, or OFF), the light is ON or 1. Only when both switches are simultaneously pressed, or contacts are opened (1 or ON), is the light OFF or 0. The action of pushing down on the switch implies a logic 1 state; releasing the switch implies a logic 0 state.

As with the AND gate, the basic operation of the 2-input NAND gate can now be expanded to include additional inputs, such as the

INPUTS		OUTPUT
A	B	Q
0	0	1
1	0	1
0	1	1
1	1	0

(a)   (b)

Fig. 2-6 Two-input NAND gate: (a) logic symbol, (b) truth table.

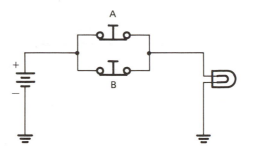

Fig. 2-7 Equivalent switch circuit of a 2-input NAND gate.

3-input NAND gate with its associated truth table shown in Figure 2–8. Since there are three inputs, each of which can either be 0 or 1, we can have one of a possible eight input combinations. From the description of the operation of the 2-input NAND gate, the output of *any* NAND gate will be 0 only if all inputs are 1. Therefore, the NAND gate has only one unique output state, that is, a logic 0 output only when all inputs are 1.

Using TTL and CMOS logic families, Table 2–4 lists the various NAND gate devices that are available.

OR Gate

The OR gate, whose logic symbol is shown in Figure 2–9, has an operation that determines if any of its inputs are at logic 1. For a 2-input OR gate:

If either input *A* OR *B* OR both are 1, then the output is 1.

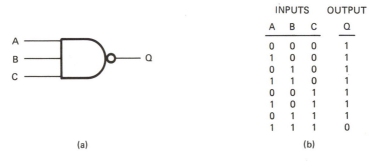

INPUTS			OUTPUT
A	B	C	Q
0	0	0	1
1	0	0	1
0	1	0	1
1	1	0	1
0	0	1	1
1	0	1	1
0	1	1	1
1	1	1	0

(a)   (b)

Fig. 2-8 Three-input NAND gate: (a) logic symbol, (b) truth table.

Table 2-4
TTL and CMOS NAND Gates

Inputs	No. of Gates per Package	TTL Series	CMOS Series
2	4	7400	4011, 74C00
3	3	7410	4023, 74C10
4	2	7420	4012, 74C20
8	1	7430	4068, 74C30
12	1	74132	None
13	1	74133	None

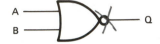

Fig. 2-9 Two-input OR gate logic symbol.

On the other hand, if both inputs are 0, then the output will be 0. We can therefore summarize the operation of a 2-input OR gate by the truth table shown in Table 2–5.

Another way of explaining the operation of a 2-input OR gate is to visualize the gate as a parallel combination of two normally open push-button switches that connect a battery to a light bulb, as shown in Figure 2–10. When both switches are released, or contacts are opened (0 or OFF), the light is OFF or 0. If either or both switches are

Table 2-5
Two-Input OR Gate
Truth Table

Input		Output
A	*B*	*Q*
0	0	0
1	0	1
0	1	1
1	1	1

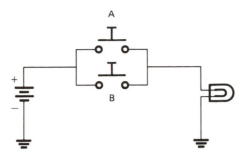

Fig. 2-10 Equivalent switch circuit of a 2-input OR gate.

simultaneously pressed, or contacts are closed (1 or ON), the light is then ON or 1.

The operation of the preceding 2-input OR gate can now be expanded to include additional inputs, such as the 3-input OR gate with its associated truth table shown in Figure 2–11. From the description of the operation of both the 2- and 3-input OR gates, we can now state that the output of *any* OR gate will be 0 only if all inputs are 0. Therefore, the OR gate has only one unique output state, that is, a logic 0 output only when all inputs are 0.

Using TTL and CMOS logic families, Table 2–6 lists the various OR gate devices that are available.

NOR Gate

The operation of the NOR gate is the opposite of the OR gate. The acronym NOR stands for NOT-OR, meaning that its operation is the NOT, or complement (i.e., opposite), of the OR gate. This additional

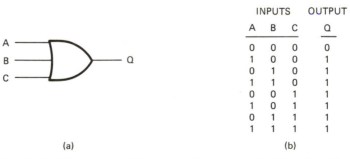

INPUTS			OUTPUT
A	B	C	Q
0	0	0	0
1	0	0	1
0	1	0	1
1	1	0	1
0	0	1	1
1	0	1	1
0	1	1	1
1	1	1	1

(a)   (b)

Fig. 2-11 Three-input OR gate: (a) logic symbol, (b) truth table.

Table 2-6
TTL and CMOS OR Gates

Inputs	No. of Gates per Package	TTL Series	CMOS Series
2	4	7432	4071, 74C32
3	3	None	4075
4	2	None	4072

feature is represented by the inversion circle added to the OR symbol. Therefore, for a 2-input NOR gate, the resulting truth table of Figure 2–12 shows that the output for the NOR gate is the opposite of the OR gate. Consequently, if both inputs A and B are 0, then the output of the 2-input NOR gate is NOT the output of the OR gate, which is then 1.

Another way of explaining the operation of a 2-input NOR gate is to visualize the gate as a series of two normally closed push-button switches that connect a battery to a light bulb, as shown in Figure 2–13. When both switches are simultaneously released (i.e., have closed contacts) (0 or OFF), the light is ON or 1. If either or both switches are then pressed (1 or ON), the light is OFF or 0.

	INPUTS	OUTPUT
	A B	Q
	0 0	1
	1 0	0
	0 1	0
	1 1	0

(a)   (b)

Fig. 2-12 Two-input NOR gate: (a) logic symbol, (b) truth table.

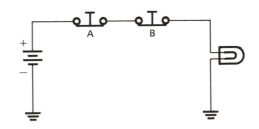

Fig. 2-13 Equivalent switch circuit of a 2-input NOR gate.

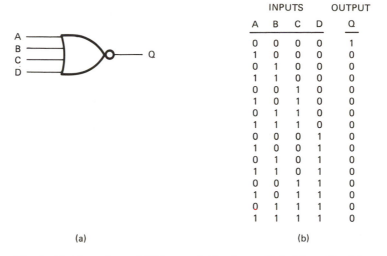

INPUTS				OUTPUT
A	B	C	D	Q
0	0	0	0	1
1	0	0	0	0
0	1	0	0	0
1	1	0	0	0
0	0	1	0	0
1	0	1	0	0
0	1	1	0	0
1	1	1	0	0
0	0	0	1	0
1	0	0	1	0
0	1	0	1	0
1	1	0	1	0
0	0	1	1	0
1	0	1	1	0
0	1	1	1	0
1	1	1	1	0

(a)   (b)

Fig. 2-14 Four-input NOR gate: (a) logic symbol, (b) truth table.

As with the OR gate, the basic operation of the 2-input NOR gate can now be expanded to include additional inputs, such as the 4-input NOR gate with its associated truth table shown in Figure 2–14. Therefore, the NOR gate has only one unique output state, that is, a logic 1 output only when all inputs are 0.

Using TTL and CMOS logic families, Table 2–7 lists the various NOR gate devices that are available.

Exclusive-OR Gate

The exclusive-OR gate (often abbreviated as XOR), whose logic symbol is shown in Figure 2–15, is a special type of 2-input OR gate that determines when its two inputs are not the same. Simply stated:

Table 2-7
TTL and CMOS NOR Gates

Inputs	No. of Gates per Package	TTL Series	CMOS Series
2	4	7402	4001, 74C02
3	3	7427	4025
4	2	None	4002
5	2	74260	None

Fig. 2-15 Exclusive-OR gate logic symbol.

If either input *A* OR *B*, but not both, is 1, then the output is 1.

On the other hand, if both inputs are simultaneously 0 or 1, then the output will be 0. We can therefore summarize the operation of a 2-input XOR gate by the truth table shown in Table 2–8. Since the XOR gate determines when the two inputs are not the same, it is sometimes called the *inequality gate*.

Using TTL and CMOS logic families, Table 2–9 lists the various XOR gate devices that are available.

Exclusive-NOR Gate

The exclusive-NOR gate (often abbreviated as XNOR), whose logic symbol is shown in Figure 2–16, is a special type of 2-input NOR gate

Table 2-8
**Two-Input XOR Gate
Truth Table**

Input		Output
A	*B*	*Q*
0	0	0
1	0	1
0	1	1
1	1	0

Table 2-9
TTL and CMOS XOR Gates

Inputs	No. of Gates per Package	TTL Series	CMOS Series
2	4	7486	4030, 4070, 74C86

Fig. 2-16 Exclusive-NOR gate logic symbol.

that determines when its two inputs are the same. In addition, its operation is the opposite of the XOR gate. Simply stated:

If either input *A* OR *B*, but not both, is 1, then the output is 0.

On the other hand, if both inputs are simultaneously 0 or 1, then the output will be 1. We can therefore summarize the operation of a 2-input XNOR gate by the truth table shown in Table 2–10. Since the XNOR gate determines when the two inputs are the same, it is sometimes called the *equality gate*.

Using TTL and CMOS logic families, Table 2–11 lists the various XNOR gate devices that are available.

Table 2-10
Two-Input XNOR Gate Truth Table

Input		Output
A	*B*	*Q*
0	0	1
1	0	0
0	1	0
1	1	1

Table 2-11
TTL and CMOS XNOR Gates

Input	No. of Gates per Package	TTL Series	CMOS Series
2	4	None	4077

ENABLING AND DISABLING GATE SIGNALS

Since AND, NAND, OR, and NOR gates each have a unique output state, each of these gates can be used to allow (enable) or inhibit (disable) digital signals through a gate.

Using a 2-input AND gate, as shown in Figure 2–17, input *A* can be used to control the passage of the digital signal present at input *B*. Whenever *A* is 0, the output *Q* is always 0 so that the AND gate is *disabled*. Whenever *A* is 1, the AND gate is enabled, and *Q* is identical to input *B*. The time relationship between input and output signals shown in Figure 2–17 is called a *timing* diagram. Figure 2–18 extends this concept using a 3-input AND gate.

Since NAND gates have the opposite operation of AND gates, Figure 2–18 shows how a 2-input NAND gate can be enabled and disabled. Enabling and disabling a NAND gate is accomplished exactly the same way as for the AND gate, except the output *Q* is now the complement of *B*, since a NAND gate is nothing more than an AND gate with an inverter.

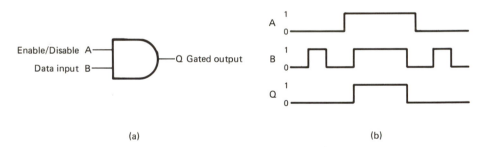

Fig. 2–17 Gating a signal using a 2-input AND gate: (a) logic diagram, (b) timing diagram.

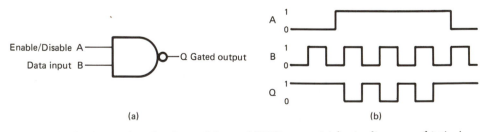

Fig. 2–18 Gating a signal using a 2-input NAND gate: (a) logic diagram, (b) timing diagram.

As shown in Figure 2–19, whenever input *A* is 1, a 2-input OR gate is disabled, and *Q* is always 1. On the other hand, whenever *A* is 0, the OR gate is enabled, and *Q* is exactly the same as *B*. Figure 2–20 illustrates the enabling and disabling of a 4-input OR gate.

Similar to the relationships between AND and NAND gates, enabling and disabling a NOR gate is accomplished exactly the same way as for the OR gate, except that the output *Q* is the complement of *B*.

GATE PROPAGATION DELAY

Associated with all logic gates is a parameter termed *gate propagation delay*, which is the time interval between the application of an input pulse and the occurrence of the resulting output pulse. That is, gate

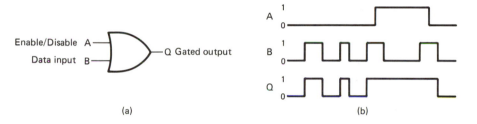

Fig. 2-19 Gating a signal using a 2-input OR gate: (a) logic diagram, (b) timing diagram.

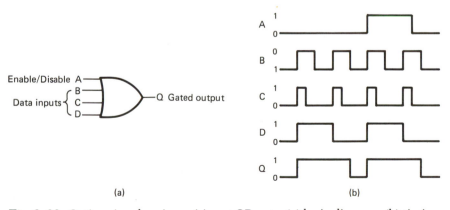

Fig. 2-20 Gating signals using a 4-input OR gate: (a) logic diagram, (b) timing diagram.

propagation delay is the time it takes for a given signal to pass through a gate.

Using an enabled AND gate, Figure 2–21 shows that two propagation delays are associated with a device: a *turn-on* delay (t_{pLH}) and a *turn-off* delay (t_{pHL}). As shown, gate propagation delay times are usually defined between the 50% points on the respective pulse edges, although other definitions may be used by various manufacturers.

As the maximum frequency of the input pulses at which the particular gate will operate is inversely related to the propagation delay through the gate, the larger the delay is, the lower the maximum frequency at which the gate will operate. In many logic gates, the turn-on and turn-off delay times are not equal. Consequently, the maximum operating speed of a gate is limited by the larger of the two propagation delays.

GATE COMBINATIONS AND SPECIAL-PURPOSE GATES

Up to this point we have discussed the inverter, AND, NAND, OR, NOR, XOR, and XNOR gates on an individual basis. In this section, we will show how to interconnect one or more of these individual logic elements to give a specified output for a given combination of input levels. This is generally referred to as *combinatorial logic.*

Inverter

If we carefully study the truth tables of both the 2-input NAND and NOR gates, we notice that, if both inputs are the same, the resultant output is the complement of the common input. Figure 2–22 shows

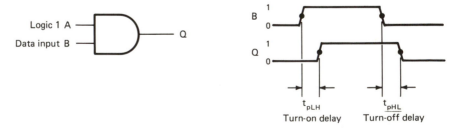

Fig. 2-21 Propagation delays through a gate.

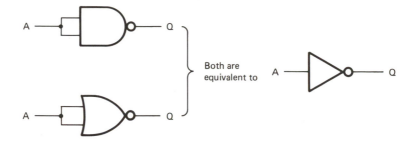

Fig. 2-22 Synthesis of an inverter with NAND and NOR gates.

how 2-input NAND and NOR gates are connected to form an inverter. In both cases, when the common input is 1, the output is 0, and vice versa. The same can also be accomplished with multi-input NAND and NOR gates.

Synthesis of Simple Gate Elements

1. AND Gate

As shown in Figure 2–23a, if we use a NAND gate followed by an inverter function, we have synthesized, or simulated, the function of an AND gate. Since we now know how to create inverters using NAND and NOR gates, we can create an AND gate from either NAND–NAND or NAND–NOR combinations.

On the other hand, we can place inverters or their equivalent at the inputs of a NOR gate to simulate an AND gate, as shown in Figure 2–24. The detailed truth table for this inverter–NOR gate combination

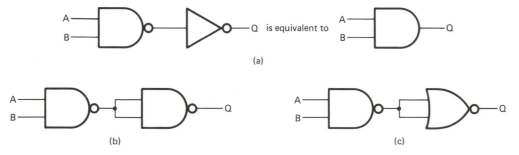

Fig. 2-23 Synthesis of a 2-input AND gate: (a) NAND–inverter, (b) NAND–NAND, (c) NAND–NOR.

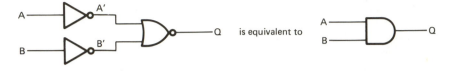

Fig. 2-24 Synthesis of a 2-input AND gate using inverters and a NOR gate.

is summarized in Table 2–12. If we want to construct a 3-input AND gate, for example, we then need a 3-input NOR gate with three inverters, and so on.

2. OR Gate

As shown in Figure 2–25, if we use a NOR gate followed by an inverter function, we have simulated the function of an OR gate. Like the AND gate simulation, an OR gate can also be had using NOR–NAND and NOR–NOR combinations.

On the other hand, we can place inverters or their equivalent at the inputs of a NAND gate to simulate an OR gate, such as the 3-input OR gate of Figure 2–26. The detailed truth table for this inverter–NAND gate combination is summarized in Table 2–13.

3. NAND and NOR Gates

AND and OR gates, as shown in Figure 2–27, can be used with inverters to simulate, respectively, NAND and NOR gates. Interestingly enough, we can construct a NOR gate solely using NAND gates, as shown in Figure 2–28, since we have already shown that we can simulate inverters, AND, and OR gates with just NAND gates. Actually, virtually *any* logic function can be built given enough NAND gates.

Table 2-12
Truth Table for Figure 2-24

	Input			Output
A	B	A'	B'	Q
0	0	1	1	0
1	0	0	1	0
0	1	1	0	0
1	1	0	0	1

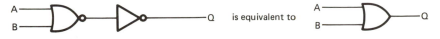

Fig. 2-25 Synthesis of a 2-input OR gate.

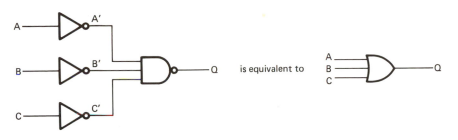

Fig. 2-26 Synthesis of a 3-input OR gate.

Table 2-13
Truth Table for Figure 2-26

Input						Output
A	*B*	*C*	*A′*	*B′*	*C′*	*Q*
0	0	0	1	1	1	0
1	0	0	0	1	1	1
0	1	0	1	0	1	1
1	1	0	0	0	0	1
0	0	1	1	1	0	1
1	0	1	0	1	0	1
0	1	1	1	0	0	1
1	1	1	0	0	0	1

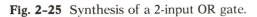

(a)

(b)

Fig. 2-27 Synthesis of NAND and NOR gates.

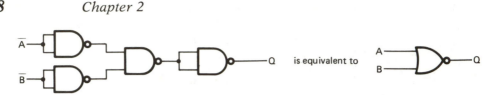

Fig. 2-28 Synthesis of a NOR gate with NAND gates.

4. XOR and XNOR Gates

Figure 2–29a shows how two inverters, two AND gates, and a single OR gate can be wired to function as a 2-input XOR gate. The detailed truth table for this combination is summarized in Table 2–14. If we add an inverter to this combination (Figure 2–29b), we then have an XNOR function. As was pointed out previously, inverters, AND, and OR gates can be built using only NAND gates. We can therefore create XOR and XNOR gates solely with NAND gates, as shown in Figure 2–30.

Specialized Combinatorial Logic

As part of their TTL and CMOS product lines, various manufacturers have developed a number of specialized combinations of logic gates. For the most part, these combinations comprise AND–OR and AND–OR–invert logic.

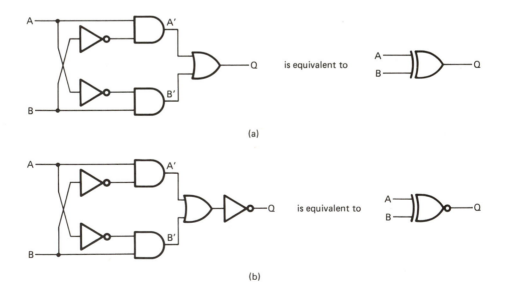

(a)

(b)

Fig. 2-29 Synthesis of (a) XOR gate, (b) XNOR gate.

Table 2-14
Truth Table for Figure 2-29

Input				Output	
A	B	A′	B′	Q (XOR)	Q (XNOR)
0	0	0	0	0	1
1	0	1	0	1	0
0	1	0	1	1	0
1	1	0	0	0	1

1. AND-OR Gates

Figure 2–31 shows the simplest AND–OR gate, which most manufacturers call a *2-wide, 2-2-input AND–OR gate*. This is because the OR gate, which is the last gate in the combinatorial chain, has two inputs (i.e., "2-wide"), and both AND gates each have two inputs (i.e., "2-2-input"). Since there are four inputs, each of which has two possible states, there are 16 possible combinations of the four inputs, as shown in the truth table of Table 2–15.

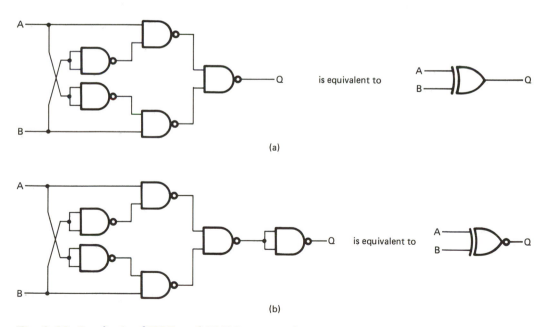

(a)

(b)

Fig. 2-30 Synthesis of XOR and XNOR gates using NAND gates.

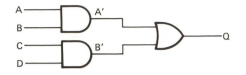

Fig. 2-31 Two-wide, 2-2-input AND–OR gate.

Table 2–16 summarizes the common TTL and CMOS AND–OR gates that are available.

2. AND–OR-Invert Gates

Figure 2–32a shows a combinatorial logic circuit consisting of two 2-input AND gates, a 2-input OR gate, and an inverter. Most

Table 2-15
Truth Table for Figure 2-31

		Input				Output
A	*B*	*C*	*D*	*A'*	*B'*	*Q*
0	0	0	0	0	0	0
1	0	0	0	0	0	0
0	1	0	0	0	0	0
1	1	0	0	1	0	1
0	0	1	0	0	0	0
1	0	1	0	0	0	0
0	1	1	0	0	0	0
1	1	1	0	1	0	1
0	0	0	1	0	0	0
1	0	0	1	0	0	0
0	1	0	1	0	0	0
1	1	0	1	1	0	1
0	0	1	1	0	1	1
1	0	1	1	0	1	1
0	1	1	1	0	1	1
1	1	1	1	1	1	1

Table 2-16
TTL and CMOS AND-OR Gates

S/N	Logic	Function
4019	CMOS	Quad 2-wide, 2-2-input AND–OR gate
7452	TTL	4-wide, 2-2-2-3-input AND–OR gate
7464	TTL	4-wide, 4-3-2-2-input AND–OR gate

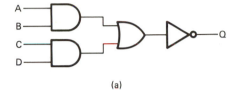   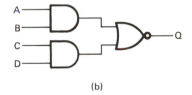

(a)   (b)

Fig. 2-32 AND–OR–invert gates.

manufacturers combine the OR gate and the inverter, as represented by a 2-input NOR gate (Figure 2–32b). The operation of this circuit is the same as the AND–OR circuit of Figure 2–31, except the output is now inverted. The common TTL and CMOS AND–OR–invert gates that are available are summarized in Table 2–17.

Transmission Gates

The CMOS transmission gate, often referred to as a *bilateral or analog switch*, is a unique device in that it has no TTL equivalent. Although it is no more a gate than is the inverter, it nevertheless deserves a brief

Table 2-17
TTL and CMOS AND-OR-Invert Gates

S/N	Logic	Function
4085	CMOS	Dual 2-wide, 2-2-input AND–OR–invert gate
7451	TTL	Dual 2-wide, 2-2-input AND–OR–invert gate
7453	TTL	4-wide, 2-2-2-2-input AND–OR–invert gate
7454	TTL	4-wide, 3-3-2-2-input AND–OR–invert gate
7455	TTL	2-wide, 4-4-input AND–OR–invert gate

introduction in this chapter because, simple as it is, it is a basic building block in the formation of CMOS flip-flops, shift registers, and counters. Digital and low-level analog signals can be switched to other circuits or buses at rates approaching 10 megahertz (MHz). Unlike a mechanical switch or relay, the transmission gate exhibits no contact bounce during switching.

Shown in Figure 2–33a, the transmission gate is formed by two pairs of p-channel and n-channel enhancement MOS transistors. One pair functions as a simple inverter, while the other pair is connected in parallel and functions as the switch. When the *control* input is at logic 1 (V_{DD}), the p- and n-channel transistors are switched ON, resulting in a low resistance path from input to output. If the control input is now at logic 0 ($-V_{SS}$), the transistors are now switched OFF, resulting in virtually an open circuit from input to output. Since the pair of p- and n-channel transistors that form the switch section is connected in parallel, *the input and output terminals are therefore interchangeable,* so the transmission gate is said to be *bilateral.* Unlike any other available logic function, digital or analog data can pass through a transmission gate in either direction, which makes them ideal for bus applications (see Chapter 8).

As integrated-circuit devices, perhaps the best known are the similar 4016 and 4066 quad transmission gates, shown by the pin diagram of Figure 2–34. When the control input is at logic 1, the ON or closed circuit resistance is rated at 300 Ω, while it is typically 80 Ω for the 4066. For a logic 0 control input, the OFF or open circuit resistance is typically 10^{11} Ω. Therefore, the signal source will not be subjected to

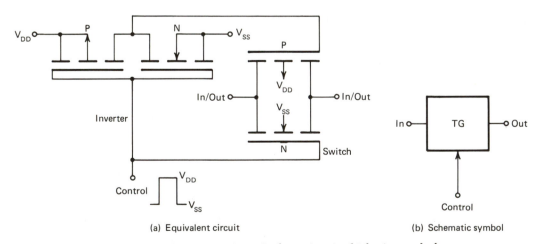

(a) Equivalent circuit   (b) Schematic symbol

Fig. 2-33 CMOS transmission gate: (a) equivalent circuit, (b) logic symbol.

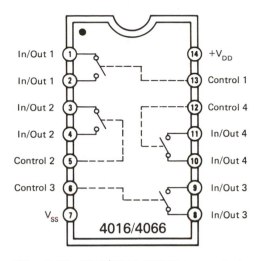

Fig. 2-34 4016/4066 CMOS transmission gate pin diagram.

loading effects when the transmission gate is OFF. The input signal range must not exceed the power supply limits. For digital signals, the input can vary from ground to +15 V; for analog signals, the input can vary from −7.5 to +7.5 V. In either case, the maximum input voltage difference $V_{DD} - V_{SS}$ is 15 V. Detailed applications for transmission gates are given in Chapter 8.

USES FOR LOGIC GATES

Besides having the ability to simulate other logic gates, certain combinations of logic elements, most notably the XOR gate, are used to form a variety of basic but useful arithmetic operations, such as adders, subtractors, and comparators.

Adders

To begin, let's define four rules for the addition of two 1-digit binary numbers, or bits:

(1) $0 + 0 = 0$

(2) $0 + 1 = 1$

(3) $1 + 0 = 1$

(4) $1 + 1 = 10$ (i.e., a decimal 2)

We can then summarize these four rules in the form of the following table:

Addend A	Augend B	Sum A + B
0	0	0
1	0	1
0	1	1
1	1	10

where the two inputs A (the *addend*) and B (the *augend*) are added together to form the *sum* of $A + B$. As shown in the last line of the table, when we add $1 + 1$, we get 10, which is not the *decimal* number 10, but rather the *binary* representation for the number 2, since we all know that $1 + 1 = 2$. Actually, the sum has two binary numbers (bits) associated with it. For the first three cases, the left or most significant bit is 0; it is omitted, as it is understood. If we were to include the most significant bit, even though it may be 0, our table for the addition of two 1-digit binary numbers can now be rewritten as

Addend A	Augend B	Sum A + B
0	0	00
1	0	01
0	1	01
1	1	10

Closer inspection of this binary addition table shows that we actually have *two* truth tables, so that

Input		Output		Equivalent Decimal Number
A	*B*	Carry	Sum	
0	0	0	0	0
1	0	0	1	1
0	1	0	1	1
1	1	1	0	2

where we have now partitioned the 2-bit number representing the sum $A + B$ into two separate columns. The most significant bit is called the *carry* output, while the rightmost or least significant bit is called the *sum* output. Consequently, the truth table for the addition of two 1-bit numbers is really represented by two truth tables: one for an output called *sum* and one for an output called *carry*.

The *sum* output has the same form as the truth table for an XOR gate, while the *carry* output has the same truth table as a 2-input AND gate. Since both outputs occur simultaneously, the truth table for the addition of two 1-bit numbers is then represented by *an XOR gate in parallel with a 2-input AND gate*, as shown in Figure 2–35a. Such a circuit is referred to as a *half-adder*.

The *full-adder* circuit, shown in Figure 2–36a, has *three* inputs and two outputs and is made from two half-adders and a 2-input OR gate. In Figure 2–36b, the equivalent logic circuit contains two XOR gates and one 2-wide 2-2-input AND–OR gate. Consequently, the basic difference between the half- and full-adder is that *the full-adder has an additional input that accepts an input carry*. The full-adder must then add the 1-bit addend (*A*) and augend (*B*) as well as the 1-bit input carry (*CI*).

In order then to add two binary numbers, each of which can be represented by more than 1 bit, we must resort to using additional full-adders. To accomplish this, we must use one half-adder for the least significant bit and a full-adder for the remaining bit, as shown by

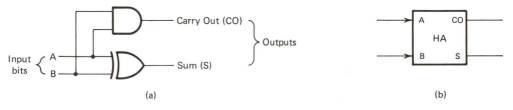

(a)   (b)

Fig. 2-35 Half-adder: (a) logic diagram, (b) logic symbol.

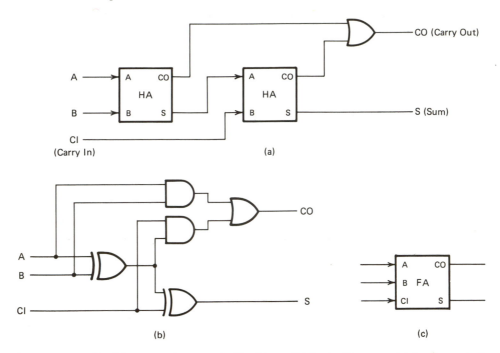

Fig. 2-36 Full-adder: (a) using two half-adders, (b) logic diagram, (c) logic symbol.

the block diagram of Figure 2–37a. The carry of the half-adder is connected to the input carry of the full-adder. A half-adder is used for the least significant bit as there is no input carry for this bit. However, a full-adder may be used in place of the half-adder if the input carry is grounded (logic 0).

For this 2-bit parallel adder, the least significant bits of two numbers to be added, A and B, are designated $A0$ and $B0$, respectively; the next higher order (most significant) bits are $A1$ and $B1$. The three sum bits are $S0$, $S1$, and $S2$, where $S2$ is the output carry from the full-adder, which in turn becomes the most significant sum bit. Figure 2–38 shows a 2-bit adder used to add the decimal numbers 3 and 1 (binary 11 and 01, respectively).

The least significant bits $A0$ and $B0$ of 3 and 1, respectively, are 1 and 1. The most significant bits $A1$ and $B1$ are 1 and 0, respectively. The truth table for a half-adder tells us that when both inputs $A0$ and $B0$ are 1, the carry (CO) is 1 and the sum ($S0$) is 0. The carry of the half-adder now becomes the input carry (CI) to the full-adder. Since $A1 = 1$, $B1 = 0$, and $CI = 1$, the truth table for a full-adder (Table 2–18) tells us that the carry of the full-adder ($S2$) is 1 while the sum ($S1$) is 0. The three sum

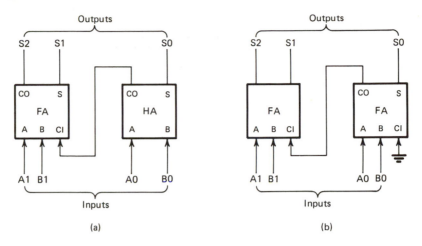

Fig. 2-37 Two-bit adder: (a) using half- and full-adders, (b) using full-adders only.

bits *S2*, *S1*, and *S0* are, in order, 1–0–0, which is the binary equivalent of the decimal number 4. This should come as no shock, as 3 + 1 always equals 4.

Using TTL and CMOS logic families, Table 2–19 lists the parallel adders currently available.

As another example, suppose we want to use a 7483 4-bit adder to add the decimal numbers *A* = 10 and *B* = 15 (binary 1010 and 1111,

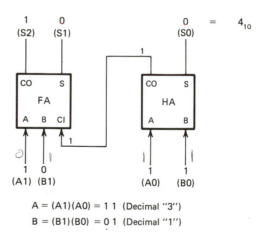

$$A = (A1)(A0) = 1\ 1 \quad (\text{Decimal ''3''})$$
$$B = (B1)(B0) = 0\ 1 \quad (\text{Decimal ''1''})$$

Fig. 2-38 Logic diagram of a 2-bit adder to add the numbers 3 and 1.

Table 2-18
Truth Table for a Full-Adder

Input			Output		Equivalent Decimal
A	B	CI	CO	S	Number
0	0	0	0	0	0
1	0	0	0	1	1
0	1	0	0	1	1
1	1	0	1	0	2
0	0	1	0	1	1
1	0	1	1	0	2
0	1	1	1	0	2
1	1	1	1	1	3

Table 2-19
TTL and CMOS Parallel Adders

No. of Bits	TTL Series	CMOS Series
2	7472	None
4	7483	4008, 74C83

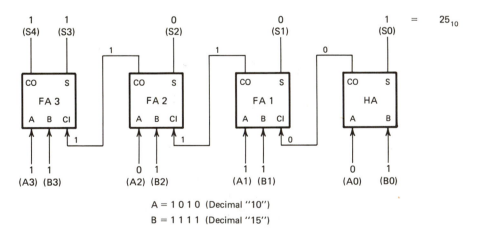

Fig. 2-39 Internal logic of a 7483 TTL 4-bit adder.

respectively). Figure 2–39 illustrates the internal block diagram of a 7483 using a half-adder and three full-adders. For the two numbers, the respective bits are

$$A0 = 0 \qquad B0 = 1$$
$$A1 = 1 \qquad B1 = 1$$
$$A2 = 0 \qquad B2 = 1$$
$$A3 = 1 \qquad B3 = 1$$

1. Half-adder: $A0 = 0$ and $B0 = 1$ give $CO = 0$ and $S0 = 1$.
2. First full-adder: $A1 = 1$, $B1 = 1$, and $CI = 0$ give $CO = 1$ and $S1 = 0$.
3. Second full-adder: $A2 = 0$, $B2 = 1$, and $CI = 1$ give $CO = 1$ and $S2 = 0$.
4. Third full-adder: $A3 = 1$, $B3 = 1$, and $CI = 1$ give $CO (S4) = 1$ and $S3 = 1$

Consequently, the five sum bits $S4$, $S3$, $S2$, $S1$, and $S0$ are 1–1–0–0–1 in binary. The equivalent decimal number can then be found by

$$2^0 + 2^3 + 2^4 = \; 1 + 8 + 16$$
$$= 25$$

which is exactly what we would expect when we add the two decimal numbers 10 and 15.

Subtractors

As with our discussion of adders, let's first define four rules for the subtraction of two 1-digit binary numbers, or bits:

(1) $0 - 0 = 0$

(2) $0 - 1 = 1$ (i.e., borrow a 1)

(3) $1 - 0 = 1$

(4) $1 - 1 = 0$

When subtracting decimal numbers, we sometimes have to *borrow* from the next or higher-order column. In binary subtraction, a borrow is required when we try to subtract 1 from 0. Under this circumstance, a binary 10 (i.e., decimal 2) is created in the column being subtracted. We can then summarize these four rules in the form of the following table:

Minuend A	Subtrahend B	Borrow BR	Difference A − B
0	0	0	0
1	0	0	1
0	1	1	1
1	1	0	0

As with the addition of two 1-bit binary numbers, closer inspection of this binary subtraction table shows that we actually have *two* truth tables, as shown in Table 2–20. Consequently, the truth table for the subtraction of two 1-bit numbers is really represented by two truth tables: one for an output called *difference* and one for an output called *borrow*.

The *difference* output has the same form as the truth table for an XOR gate, while the *borrow* output has the truth table for some kind of 2-input gate, which can be represented by a 2-input AND gate with a single inverted input. Since both outputs simultaneously occur, the truth table for the subtraction of two 1-bit numbers is then represented by *an XOR gate in parallel with a 2-input gate*, as shown in Figure 2–40a. Such a circuit is referred to as a *half-subtractor*.

The *full-subtractor* circuit shown in Figure 2–41a, like the full-adder, has *three* inputs and two outputs and is made from two half-subtractors and a 2-input OR gate. In Figure 2-41b, the equivalent logic circuit contains two XOR gates, two inverters, and one 2-wide 2-2-input AND–OR gate. Consequently, the basic difference between the half- and full-subtractor is that *the full-subtractor has an additional input that accepts an input borrow*. Table 2–21 is the truth table for a full-subtractor.

Table 2-20
Truth Tables for the Subtraction of 1-Bit Numbers

Input		Output		Input		Output
A	B	Borrow		A	B	Difference
0	0	0		0	0	0
1	0	0	*and*	1	0	1
0	1	1		0	1	1
1	1	0		1	1	0

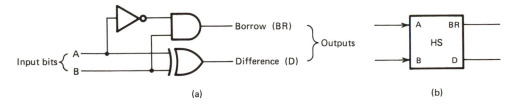

Fig. 2-40 Half-subtractor: (a) logic diagram, (b) logic symbol.

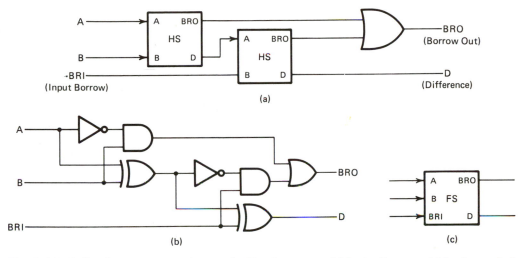

Fig. 2-41 Full-subtractor: (a) using two half-subtractors, (b) logic diagram, (c) logic symbol.

Table 2-21
Truth Table for a Full-Subtractor

Input			Output	
A	*B*	*BRI*	*BRO*	*D*
0	0	0	0	0
1	0	0	0	1
0	1	0	1	1
1	1	0	0	0
0	0	1	1	1
1	0	1	0	0
0	1	1	1	0
1	1	1	1	1

Comparators

The simplest comparators are the XOR and XNOR gates. From Table 2–8, the output for an XOR gate is 0 if *both* inputs simultaneously are 0 or 1. On the other hand, if both inputs are not the same, the output is 1. Since the output of an XNOR gate is the complement of the XOR gate, its output is 1 when both inputs are simultaneously the same. In effect, both the XOR and XNOR gates are 1-bit comparators.

To compare numbers having 2 bits each, we must have an additional XOR (or XNOR) gate, as shown in Figure 2–42a. Using XOR gates, the least significant bits (LSBs), $A0$ and $B0$, of the 2-bit numbers A and B are compared by XOR gate $G0$, while the two remaining bits $A1$ and $B1$ (i.e., the most significant bits, MSBs) are compared by XOR gate $G1$. If either $A0$ and $B0$ or $A1$ and $B1$ are not equal, a 1 is present on the output of the corresponding XOR gate. The 2-input OR gate is then used to determine if, bit for bit, the numbers A and B are equal. Only if $A0 = B0$ *and* $A1 = B1$ is the output of the OR gate a 0. When the numbers are not equal, a 1 appears at the output of one or both XOR gates, causing the output of the OR gate to be 1. Using XNOR gates in conjunction with an AND gate (Figure 2–42b), we have the opposite effect: if both numbers, bit for bit, are equal, then the output of the AND gate will be a 1. Table 2–22 summarizes the operation of a 2-bit comparator using XOR and XNOR gates. Consequently, this scheme may be extended to compare higher multibit numbers.

Although this method tells us whether or not two numbers A and B are equal, we nevertheless have no indication as to which number is larger. When comparing two numbers, we should have some indication if A is less than, equal to, or greater than B. Figure 2–43 shows a method of comparing two 2-bit numbers and generating an $A < B$, an $A = B$, or an $A > B$ output. From Table 2–23 we see that at input conditions 1, 4, 13, and 16 the numbers A and B are equal ($A0 = B0$ *and* $A1 = B1$). The output of the $A = B$ AND gate $G3$ is 1, while both the $A > B$ and $A > B$ OR gates are 0. When $A > B$, we have input conditions 2, 5, 6, 7, 8, and 14, so

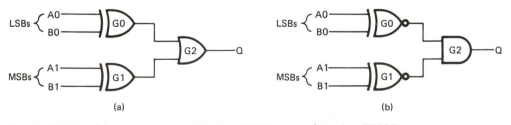

Fig. 2-42 Two-bit comparator: (a) using XOR gates, (b) using XNOR gates.

Table 2-22
Truth Table for a 2-Bit Comparator

Input				Output	
A0	*B0*	*A1*	*B1*	XOR–OR	XNOR–AND
0	0	0	0	0	1
1	0	0	0	1	0
0	1	0	0	1	0
1	1	0	0	1	0
0	0	1	0	1	0
1	0	1	0	1	0
0	1	1	0	1	0
1	1	1	0	1	0
0	0	0	1	1	0
1	0	0	1	1	0
0	1	0	1	1	0
1	1	0	1	1	0
0	0	1	1	1	0
1	0	1	1	1	0
0	1	1	1	1	0
1	1	1	1	0	1

OR gate $G6$ is now 1 while AND gate $G3$ and OR gate $G9$ are 0. Finally, when $A < B$, we have input conditions 3, 9, 10, 11, 12, and 15, so $G9$ is 1, while $G3$ and $G6$ are 0.

Using TTL or CMOS devices, there is presently only a 4-bit *magnitude comparator*, whose block diagram is given in Figure 2–44. The TTL version is the 7485, while equivalent CMOS devices have the following numbers: 74C85, 4063, 340085, and MC14585, depending on the manufacturer. For this magnitude comparator, we have basically the same inputs and outputs as for the 2-bit comparator circuit of Figure 2–43, except that we now have three *cascading* inputs. These three additional inputs allow several 4-bit comparators to be *cascaded*, or *connected in series*, for the comparison of numbers having more than 4 bits each. To compare two 16-bit numbers, as illustrated in

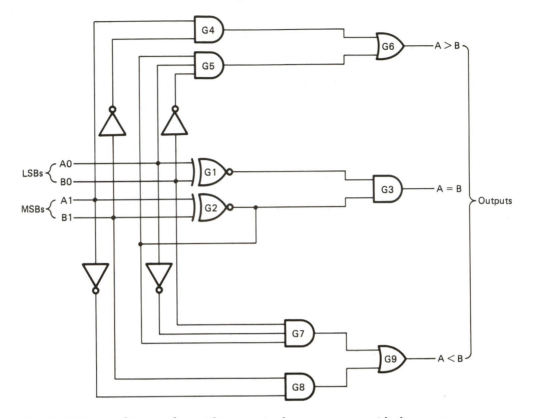

Fig. 2-43 Logic diagram for a 4-bit magnitude comparator with three outputs.

Figure 2–45, the $A > B$, $A = $ B, and $A < B$ outputs from the least significant 4-bit comparator are connected to the corresponding cascading inputs of the next higher comparator, and so on. In addition, for the least significant comparator, the $A = $ B input is set at logic 1, while the $A > B$ and $A < B$ inputs are set at 0.

REVIEW

In this chapter we have covered quite a bit of ground in discussing the basic building block of all digital logic, the gate. In addition to discussing each gate function by itself, we also saw how certain gate functions may be built from other gate functions and how we can combine gates to create adders, subtractors, and comparators. In

Table 2-23
Truth Table for an Improved 2-Bit Comparator

Input Condition	Input				Output		
	$A0$	$B0$	$A1$	$B1$	$A > B$	$A = B$	$A < B$
1	0	0	0	0	0	1	0
2	1	0	0	0	1	0	0
3	0	1	0	0	0	0	1
4	1	1	0	0	0	1	0
5	0	0	1	0	1	0	0
6	1	0	1	0	1	0	0
7	0	1	1	0	1	0	0
8	1	1	1	0	1	0	0
9	0	0	0	1	0	0	1
10	1	0	0	1	0	0	1
11	0	1	0	1	0	0	1
12	1	1	0	1	0	0	1
13	0	0	1	1	0	1	0
14	1	0	1	1	1	0	0
15	0	1	1	1	0	0	1
16	1	1	1	1	0	1	0

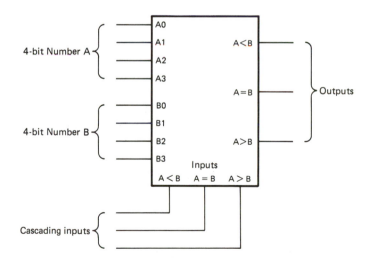

Fig. 2-44 Schematic symbol for a 4-bit magnitude comparator.

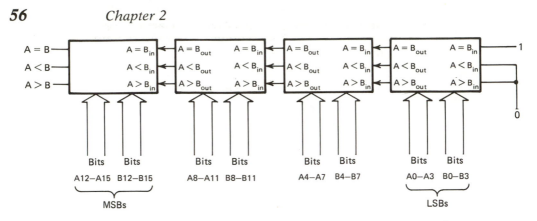

Fig. 2-45 Sixteen-bit comparator using magnitude comparators in cascade.

virtually all these cases, there exist TTL and CMOS devices that functionally accomplish these tasks with less physical space and cost.

INTRODUCTION TO THE EXPERIMENTS

The following experiments are designed to demonstrate the operation of gates and how gates can enable and disable signals. The experiments that you will perform can be summarized as follows:

Experiment	Purpose
2–1	Demonstrates the input and output relationships of 2-input AND, OR, NAND, NOR, and XOR gates by constructing their associated truth tables.
2–2	Demonstrates how gate signals may be enabled and disabled using 2-input AND, OR, NAND, NOR, and XOR gates.
2–3	Demonstrates the operation of a 4-bit parallel adder, using a 7483 TTL device.

EXPERIMENT 2-1

PURPOSE

The purpose of this experiment is to demonstrate the input and output relationships of 2-input AND, OR, NAND, NOR, and XOR gates by constructing their associated truth tables.

PIN CONFIGURATION
OF INTEGRATED-CIRCUIT DEVICES

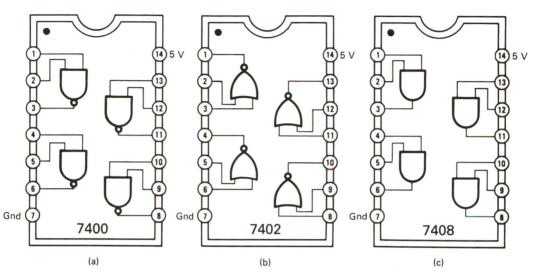

(a)   (b)   (c)

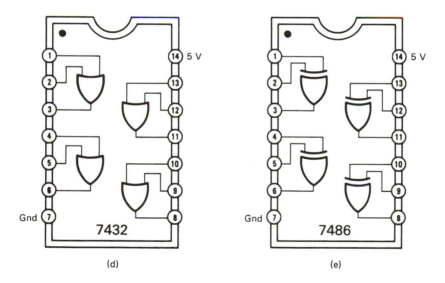

(d)   (e)

Fig. 2-46 7400, 7402, 7408, 7432, and 7486 pin diagrams.

Required Parts and Equipment

1   7400 TTL quad 2-input NAND gate
1   7402 TTL quad 2-input NOR gate
1   7408 TTL quad 2-input AND gate
1   7432 TTL quad 2-input OR gate
1   7486 TTL quad XOR gate
2   TTL logic switches
5   TTL LED monitors
1   solderless breadboard
1   TTL power supply (5 V)

SCHEMATIC DIAGRAM OF CIRCUIT

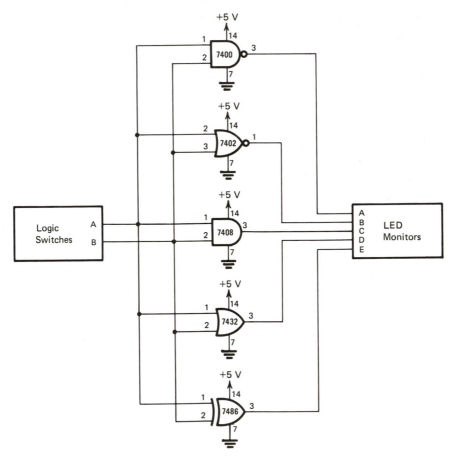

Fig. 2-47 Schematic diagram for Experiment 2–1.

STEP 1

Wire the circuit shown in the schematic diagram. In general, *this diagram will be the only one where the specific power supply connections to any integrated circuit device will be shown, as they are normally assumed to be required and are generally omitted from the diagram.* Although each integrated-circuit device in this experiment contains four equivalent but separately wired gates with a common power supply connection, only one gate from each package will be used. LED monitors are wired such that they indicate a logic 1 state when lit.

STEP 2

Apply power to the breadboard and set both input logic switches to logic 0. If you have wired everything correctly, the output Q of the five gates should be as follows:

7400 NAND gate	= 1	(LED monitor A)
7402 NOR gate	= 1	(LED monitor B)
7408 AND gate	= 0	(LED monitor C)
7432 OR gate	= 0	(LED monitor D)
7486 XOR gate	= 0	(LED monitor E)

If not, check your wiring before going further. Record your results as a truth table in Table 2–24.

Table 2-24
Data for Experiment 2-1

Input		Output Q				
A	B	NAND	NOR	AND	OR	XOR
0	0					
1	0					
0	1					
1	1					

STEP 3

Now vary the settings of the two logic switches according to Table 2–24 and record the output state for all five gates at each setting.

STEP 4

From the truth table, what do you notice about the outputs of the NAND and AND gates? Of the NOR and OR gates? Of the XOR gate?

The outputs of the NAND and AND gates are *complementary;* that is, they are the opposite of each other. The output of an AND gate is 1 only when all its inputs are 1, while the output of a NAND (i.e., NOT-AND) gate is 0 only when all its inputs are 1. Likewise, the outputs of the NOR and OR gates are complementary. The output of an OR gate is 0 only when all its inputs are 0, while the output of a NOR (i.e., NOT-OR) gate is 1 when all its inputs are 0.

The output of an XOR gate is 0 only when both its inputs are equal, either both 0 or both 1. If you performed everything correctly, you should find that not one of the five gates behaves exactly the same as any of the remaining four types.

STEP 5

If you had an 8-input NAND gate, such as the 7430 TTL version, what would be its output if any of its inputs were 0? When would its output be 0?

The output of the NAND gate will always be 1 if any of its inputs are 0, and will be 0 only when *all* inputs are 1.

EXPERIMENT 2-2

PURPOSE

Demonstrates how gate signals may be enabled and disabled using 2-input AND, OR, NAND, NOR, and XOR gates.

PIN CONFIGURATION
OF INTEGRATED-CIRCUIT DEVICES

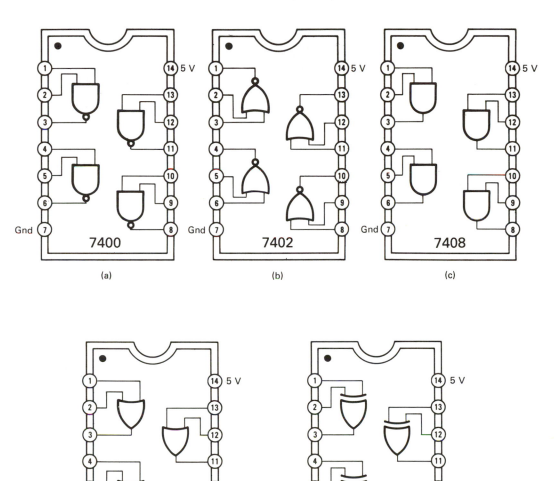

Fig. 2-48 7400, 7402, 7408, 7432, and 7486 pin diagrams.

Required Parts and Equipment

1   7400 TTL quad 2-input NAND gate
1   7402 TTL quad 2-input NOR gate
1   7408 TTL quad 2-input AND gate
1   7432 TTL quad 2-input OR gate
1   7486 TTL quad XOR gate
1   TTL logic pulser
2   TTL LED monitors
1   TTL square-wave clock (approximately 1 Hz)
1   solderless breadboard
1   TTL power supply (5 V)

SCHEMATIC DIAGRAM OF CIRCUIT

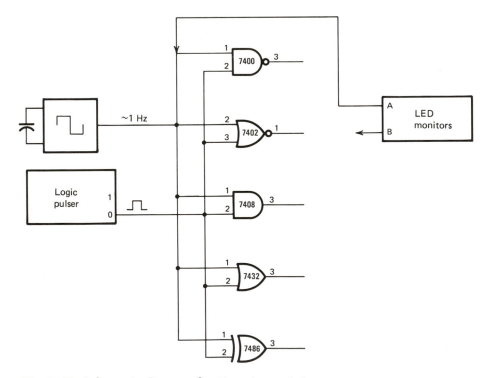

Fig. 2-49 Schematic diagram for Experiment 2–2.

STEP 1

Wire the circuit shown in the schematic diagram. Don't forget the 5-V power supply connections for each device, as they are not shown on the diagram. Make sure that the pulser output is normally at logic 0.

STEP 2

Apply power to the breadboard and adjust the rate of the clock so that LED monitor A flashes on and off approximately once each second, or slow enough for your eye to follow. With the pulser output at logic 0, sequentially check the output of each of the five gate types with LED monitor B and compare its status with that of LED monitor A (i.e, always logic 0, always logic 1, same as or the complement of LED monitor A). Record your results for each gate in Table 2–25.

STEP 3

Now press the pulser so that its output is at logic 1. As in step 2, sequentially check the output of each of the five gate types with LED monitor B and compare its status with that of LED monitor A. Record your results for each gate in Table 2–25.

Of these five gates, which one does not "gate" or inhibit the train of clock pulses from passing through?

The only gate that does not completely inhibit the input clock signal is the type 7486 XOR gate. Depending on the state of the pulser, the XOR gate either passes the clock signal through unchanged (pulser at logic 0) or inverts the clock signal (pulser at

Table 2-25
Data for Experiment 2-2

Pulser	Output Q				
Status	NAND	NOR	AND	OR	XOR
Logic 0					
Logic 1					

Use the following codes:

    0 = gate output always logic 0
    1 = gate output always logic 1
    S = gate output the same as clock input
    C = gate output is the complement of clock input

logic 1). The other four gates have their outputs either always at logic 0 or 1 when the pulser is in one position, while the gate output is either the same as or the complement of the clock input signal when the pulser is in the opposite position. It should be noted that no two gates behave the same!

EXPERIMENT 2-3

PURPOSE

The purpose of this experiment is to demonstrate the operation of a 4-bit parallel adder, using a 7483 TTL device.

**PIN CONFIGURATION OF
INTEGRATED-CIRCUIT DEVICE**

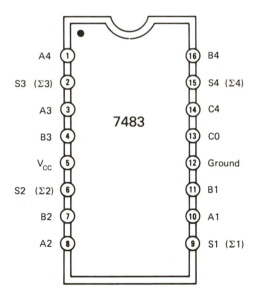

Fig. 2-50 7483 pin diagram.

Required Parts and Equipment

1 7483 TTL 4-bit parallel adder
8 TTL logic switches
5 TTL LED monitors
1 solderless breadboard
1 TTL power supply (5 V)

SCHEMATIC DIAGRAM OF CIRCUIT

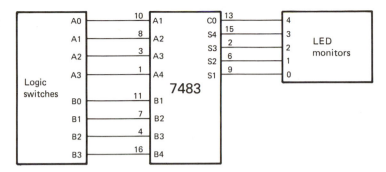

Fig. 2-51 Schematic diagram for Experiment 2–3.

STEP 1

Wire the circuit shown in the schematic diagram. If you do not have enough logic switches, use wire connected at each input either to ground (logic 0) or +5 V (logic 1), as appropriate, in place of the logic switches.

STEP 2

Apply power to the breadboard and set all logic switches to 0. Consequently, Number *A* (the addend) and Number *B* (the augend) are both 0. If you have wired everything correctly, all five LED monitors should indicate a logic 0. This makes sense, since $0 + 0 = 0$.

STEP 3

Now set the addend logic switches *A*0 to *A*3 equal to 1010, respectively (i.e., the decimal number 10) and the augend logic switches *B*0 to *B*3 equal to 1111 (i.e., the decimal number 15). Consequently, these eight inputs are

$$A0 = 0 \qquad B0 = 1$$
$$A1 = 1 \qquad B1 = 1$$
$$A2 = 0 \qquad B2 = 1$$
$$A3 = 1 \qquad B3 = 1$$

What is the status of the five LED monitors, and what is the equivalent decimal number?

You should find that LED monitors 0 to 4, representing the sum of $A + B$, should indicate the logic states 1, 1, 0, 0, and 1, respectively, which is equivalent to the decimal number 25, since

$$2^0 + 0 + 0 + 2^3 + 2^4 = 1 + 8 + 16$$
$$= 25$$

Record your results in Table 2–26.

STEP 4

Now set the eight logic switches as indicated in Table 2–26 and record your result for each sum in addition to filling in the equivalent decimal numbers. You should be able to verify that the status of the LED monitors indicates the sum of the numbers A and B, both as binary numbers as well as their decimal equivalents.

Table 2-26
Data for Experiment 2-4

	Addend A					Augend B					Sum $A + B$				
A3	A2	A1	A0	Decimal Number	B3	B2	B1	B0	Decimal Number	S4	S3	S2	S1	S0	Decimal Number
1	0	1	0		1	1	1	1							
0	1	0	1		0	1	0	0							
0	1	1	1		0	1	1	1							
1	0	0	0		0	1	0	1							
1	0	0	1		0	1	1	0							
1	1	1	0		0	1	1	1							
1	0	1	1		0	0	0	0							
0	1	0	0		0	0	1	0							
1	1	1	1		1	1	1	0							

CLOCKED LOGIC: LATCHES AND FLIP-FLOPS

OBJECTIVES

After completing this chapter, you will be able to understand:

- *The operation of R–S and D-type latches using either NOR or NAND gates.*
- *How to form a debounced switch.*
- *How to specify pull-up and pull-down resistors.*
- *The operation of R–S, J–K, and D-type flip-flops.*

SIMPLE LATCHES

A *latch* is a logic function whose output remains at a particular logic level after it has been forced to that state by an external signal. In addition, the output of a latch will remain at that level even though the external signal that forced this action may no longer be present. Consequently, a latch *remembers* what state it is in. In this section, the two basic types of latches are discussed: the *R–S* latch and *D*-type latch.

R–S *Latch*

The simplest latch is called the *R–S* or *set–reset* latch and is built using a pair of 2-input NOR gates, as shown in Figure 3–1. As shown in the logic diagram, the *R–S* latch has a S (set) and a R (reset) input and two complementary outputs Q and $\overline{Q}$. Although the letters R and S stand for "reset" and "set," this latch circuit is written as *R–S* latch, but pronounced *set–reset*. In either case, both NOR gates are said to be *cross-coupled*, as the output of each gate is connected to an input of the other gate. Therefore, the output of one gate is *fed back* to one input of the remaining gate.

To understand the operation of the *R–S* latch, let's first assume that both the R and S inputs are initially 0 and the $\overline{Q}$ output ($G1$) is 1, as illustrated in Figure 3–1a. This frequently happens when power is first applied to a latch. Consequently, since the R input is 0, the remaining input of $G2$ is 1 (it is connected to the output of $G1$), so the output of $G2$ must be 0. Since Q and $\overline{Q}$ are complementary and since $\overline{Q} = 1$, then Q must be 0, which we have just shown to be the case. Furthermore, since $Q = 0$ (which is connected as an input to $G1$) and $S = 0$, this ensures that the output of $G1$ is 1.

As shown in Figure 3–1b, by now having $S = 1$, then $\overline{Q}$ must be 0 while $Q = 1$. When $Q = 1$, the latch is said to be *set*. Furthermore, the latch will remain set indefinitely even after the set input is returned to 0 (Figure 3–1c). Now, when a logic 1 is applied to the reset input, the output of $G2$ is then 0, which implies that $\overline{Q} = 1$ (Figure 3–1d). When a logic 1 is applied to the R input so that $Q = 0$, the latch is *reset* and will remain in this state even if the reset input is then returned to logic 0 (Figure 3–1e). Thus

1. To set latch, $Q = 1$.

2. To reset latch, $Q = 0$.

When reset, the NOR gate latch remains there indefinitely until a logic 1 is applied to the set input. In other words, the latch "remembers" what state it is in until it is intentionally changed.

Table 3–1 is the truth table for the NOR-style *R–S* latch. A condition that is not allowed for this latch is when both the R and S inputs are 1. When both inputs are simultaneously 1, then both outputs will try to be 0, which violates the complementary output relationship. The output notations $Q(t-1)$ and $\overline{Q}(t-1)$ mean that the particular output will be the same as the output was previously. In other words,

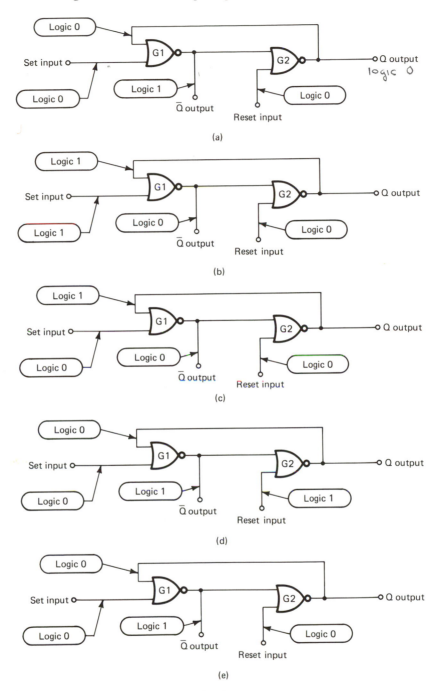

Fig. 3-1 Operation of the *R–S* latch.

Table 3-1
R–S NOR Gate Latch Truth Table

Input		Output		
S	R	Q	$\overline{Q}$	Remarks
0	0	$Q(t-1)$	$\overline{Q}(t-1)$	No change; remains in present state
1	0	1	0	Sets the latch
0	1	0	1	Resets the latch
1	1	?	?	Disallowed; output not predictable

there is no change. At the present time, there are no *R–S* NOR-gate-style latches manufactured as a single integrated circuit.

The arrangement of the two NOR gates, as shown in Figure 3–1, has been drawn in this manner to facilitate the discussion. As a general practice, the NOR gate *R–S* latch is usually drawn in the arrangement of Figure 3–2a.

A NAND-gate-style *R–S* latch has an operation that is opposite to the NOR-style latch (Figure 3–3); it is sometimes called an *S–R latch.* That is, the condition where both inputs are 0 should be avoided. The truth table for the NAND style or *S–R* latch is given in Table 3–2. The NAND gate, or *S–R*, latch is usually drawn as shown in Figure 3–4a.

The only available *S–R* latch package is the 74279 TTL *S–R* latch, which has four identical cross-coupled NAND gates in a single package, as shown by the pin diagram of Figure 3–5.

The set–reset latch enjoys widespread application as a *debounced* switch, as shown in Figure 3–6. The contacts of a switch actually bounce several times as they make contact. In terms of digital logic, a "bouncy" switch gives the appearance of an ideal switch that has just been opened and closed several times in succession. Since we ideally

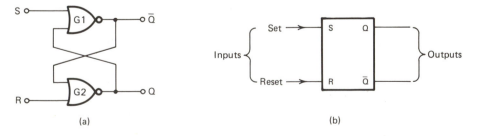

(a)　　　　　　　　　　　　　　(b)

Fig. 3-2 NOR gate *R–S* latch: (a) logic diagram, (b) logic symbol.

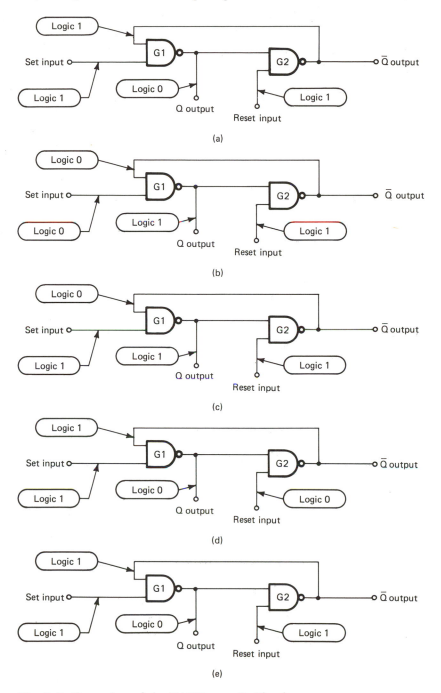

Fig. 3-3 Operation of the NAND gate *R–S* latch.

Table 3-2
***R-S* NAND Gate Latch Truth Table**

| Input | | Output | | |
S	R	Q	$\overline{Q}$	Remarks
1	1	$Q(t-1)$	$\overline{Q}(t-1)$	No change; remains in present state
0	1	1	0	Sets the latch
1	0	0	1	Resets the latch
0	0	?(1)	?(1)	Disallowed; output not predictable

want a switch that makes one and only one contact each time it is either opened or closed, we must then condition or *debounce* the switch contacts before they can become useful for digital applications.

The *switch debouncer* uses a pair of 2-input NAND gates, a mechanical SPDT switch, and a pair of resistors. A resistor, called a *pull-up* resistor, is connected to each input and the positive supply voltage, or logic 1, to give a logic 1 to the input that is not connected to the grounded *pole* or wiper contact of the mechanical SPDT switch. Since an input to any TTL or CMOS gate element should never be left *floating*, these pull-up resistors make sure that there is a logic 1 level at the input that is not connected to ground, or logic 0.

The value of the pull-up resistor is determined from

$$R = \frac{V_{IH} \min}{I} \qquad (3-1)$$

where

$$V_{IH} \min = \text{minimum input voltage required to produce a logic 1}$$

$$I = \text{input current}$$

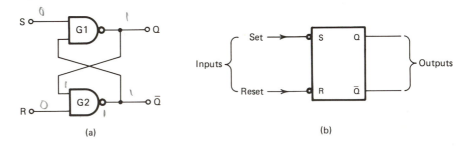

(a)   (b)

Fig. 3-4 NAND gate *R–S* latch: (a) logic diagram, (b) logic symbol.

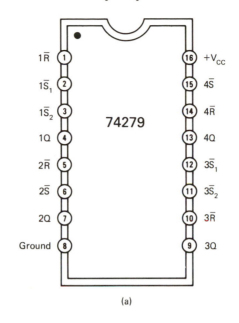

(a)

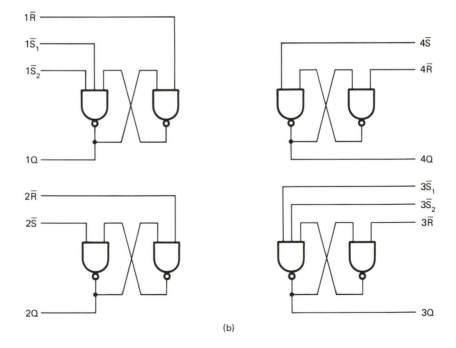

(b)

Fig. 3-5 74279 TTL quad *S–R* latch pin diagram.

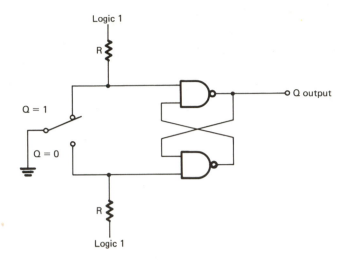

Fig. 3-6 Debounced switch contacts using cross-coupled NAND gates.

For TTL NAND gates, V_{IH} min is typically 2 V, although 2.4 V is usually taken as an average (see Chapter 8). Consequently, 2.6 to 3 V is dropped across the pull-up resistor. The current drawn by a typical TTL gate input is 1.6 milliamperes (mA), so suitable pull-up resistor values are in the 1.6 to 1.8-kΩ range. For CMOS NAND gates, the pull-up resistor is typically 22 kΩ for supply voltages from 5 to 15 V.

On the other hand, a NOR-gate-style latch can be used as a debounced switch as shown in Figure 3–7. In this case, the wiper of the switch is tied to logic 1, and *pull-down* resistors are connected from

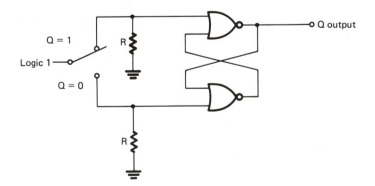

Fig. 3-7 Debounced switch contacts using cross-coupled NOR gates.

each input to ground. The value of the pull-down resistor is determined from

$$R = \frac{V_{IL}\max}{I} \qquad (3\text{--}2)$$

where

$$V_{IL} \max = \text{maximum input voltage required to produce a logic 0}$$

$$I = \text{input current}$$

For TTL NOR gates, V_{IL} max is typically 0.8 V, so 0.8 V is dropped across the pull-down resistor. As with the NAND gate, the current drawn by a gate input is 1.6 mA, so the maximum pull-down resistor value is 500 Ω, although a value of 330 Ω is generally used. For CMOS NAND gates, the pull-down resistor is typically 22 kΩ.

D-*Type Latch*

As we have pointed out for both types of *R–S* latches, there is a condition that should be avoided. For the NOR gate latch, both inputs cannot be at logic 1 simultaneously; for the NAND gate latch, both inputs cannot be logic 0. One way of avoiding this possible situation is to use a single inverter with a *R–S* latch, an arrangement called a *D*- or *data*-type latch.

　　Figure 3–8 shows a *D*-type latch using NOR gates and an inverter. The inverter makes it impossible for both inputs to be the same logic level at any given time, as summarized by the truth table of Table 3–3.

　　On the other hand, *D*-type latches are also possible using NAND gates, as shown in Figures 3–9 and 3–10. The arrangement of Figure 3–9

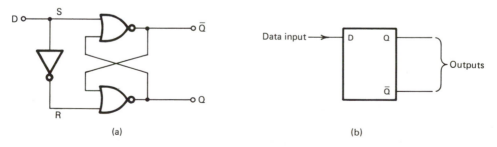

(a)　　　　　　　　　　　　　　　　(b)

Fig. 3-8 *D*-type latch using NOR gates: (a) logic diagram, (b) logic symbol.

Table 3-3
Truth Table for *D*-Type Latch Using NOR Gates

Input			Output		
D	*S*	*R*	*Q*	$\overline{Q}$	Remarks
—	0	0	—	—	Not possible
1	1	0	1	0	Sets the latch
0	0	1	0	1	Resets the latch
—	1	1	—	—	Not possible

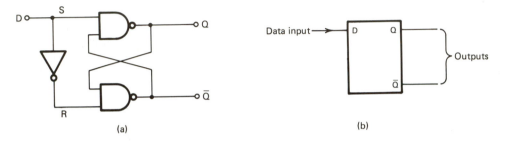

Fig. 3-9 *D*-type latch using NAND gates having active HIGH inputs: (a) logic diagram, (b) logic symbol.

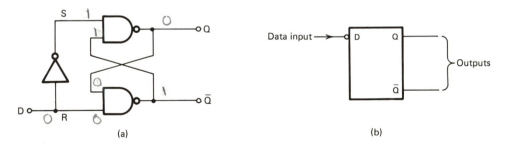

Fig. 3-10 *D*-type latch using NOR gates having active LOW inputs: (a) logic diagram, (b) logic symbol.

has an *active-HIGH* input, so $Q = 1$ when $D = 1$, as given by Table 3–4. Figure 3–10 shows the *active-LOW* variation of the D-type NAND gate latch; here $Q = 1$ when $D = 0$.

The TTL version of a *D*-type latch is the 7475, as shown in Figure 3–11a, where four latches are contained in a single package. Notice that

Table 3–4
Truth Table for Active-HIGH *D*-Type Latch Using NAND Gates

Input			Output		
D	*S*	*R*	*Q*	$\overline{Q}$	Remarks
—	0	0	—	—	Not possible
0	0	1	0	1	Sets the latch
1	1	0	1	0	Resets the latch
—	1	1	—	—	Not possible

two latches share a common gate input, which is called the *enable* input. The available CMOS equivalents are the 4042 and 74C75 (Figure 3–11b and c). The 74C77 CMOS device (Figure 3–11d) is the same as the 74C75 except that the $\overline{Q}$ outputs are not provided.

FLIP-FLOPS

A flip-flop is a logic function whose output can have two stable states (0 or 1), plus the ability to change states when a control signal is applied. Furthermore, the flip-flop remains in that state even after the input signals are removed. Consequently, flip-flops, like latches, are considered as simple memory elements, as they have the ability to remember the state of their outputs when the inputs are removed. Since there are two stable states, flip-flops and latches are often referred to as *bistable elements*. Basically, latches and flip-flops perform the same function; they differ primarily in the manner in which they are triggered, or activated.

As discussed in the previous section, the latch can change state as long as the gate or enable input is 1. Consequently, the latch is said to be controlled by the *level* of a certain input, which in this case is the gate input. On the other hand, flip-flops are controlled by the *transition* of the triggering signal, which is referred to as the *clock* input. That is, the output of the flip-flop can be made to change states simultaneously when the clock input makes the transition from 0 to 1 (or 1 to 0). Since flip-flops are primarily triggered by the transition, which may be either the leading or trailing edge of the clock signal, they are frequently referred to as being *edge-triggered* flip-flops.

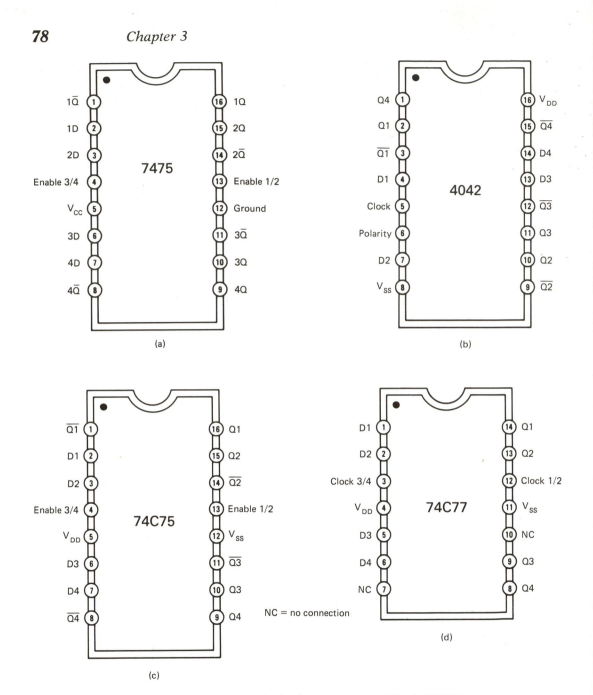

Fig. 3-11 Integrated-circuit latches: (a) 7475, (b) 4042, (c) 74C75, (d) 74C77.

The remainder of this chapter discusses the three most frequently encountered types of flip-flops and their variations: the *R–S*, *J–K*, and *D*-types.

R-S *Flip-Flop*

From the basic *R–S* latch, it is possible to create an improved latch circuit. One such improvement is the *R–S* flip-flop circuit of Figure 3–12. With the addition of two NAND gates, called *steering gates*, we now have three inputs: *S* (set), *R* (reset), and *G* (gate). The gate input is also called the *clock (CLK or CK)* input. The operation of this gated latch is the same as the basic *R–S* NAND gate latch, except that *no output changes will occur unless the gate input is logic 1*. The timing diagram of Figure 3–13 illustrates this operation.

Instead of requiring external set and reset inputs, the two outputs of an *R–S* flip-flop can be fed back cross-coupled style to the *S* and *R* inputs, as shown in Figure 3–14. Output $\overline{Q}$ is connected to the *S* input, while *Q* is connected to *R*. This single input arrangement is referred to as a *type T*, or *toggle*, flip-flop. From the timing diagram, it takes two gate or clock pulses to cause the *Q* output to repeat a 0 to 1 transition. For this reason, the *T*-type flip-flop is often referred to as a *divide-by-2 counter* or *binary divider*. More about binary dividers and divide-by-*N* counters is given in Chapter 5.

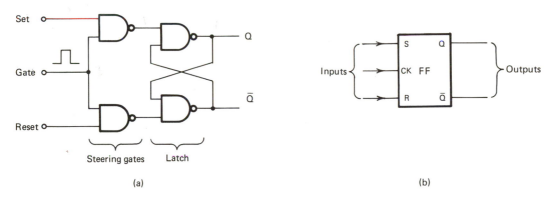

Fig. 3–12 *R–S* flip-flop: (a) logic diagram, (b) logic symbol.

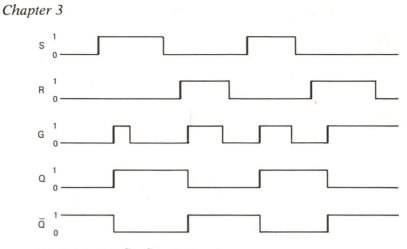

Fig. 3-13 *R–S* flip-flop timing diagram.

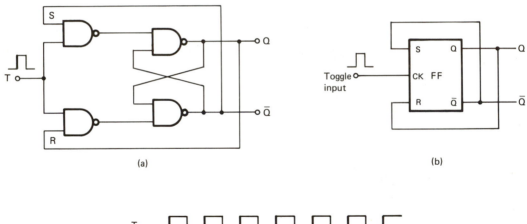

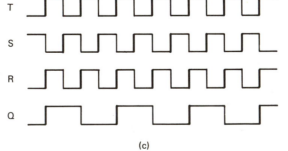

Fig. 3-14 *T*-type flip-flop: (a) logic diagram, (b) logic symbol, (c) timing diagram.

J-K *Flip-Flop*

Of all the flip-flop configurations, the *J–K flip-flop* is perhaps the most widely used. The letters *J* and *K* have no real significance other than being two consecutive letters of the alphabet. In this section, two types of *J–K* flip-flops are discussed: the basic edge-triggered *J–K* flip-flop and the master–slave *J–K* flip-flop.

1. Edge-Triggered *J-K* Flip-Flop

The basic *J–K* edge-triggered flip-flop is shown in Figure 3–15. If both the *J* and *K* inputs are at logic 0, there is no change in either output after the *CK* input goes from logic 1 to 0. If *J* = 1 and *K* = 0, output *Q* either stays at or goes to logic 1 so that *Q* = *J* after the *CK* input goes from 1 to 0. If *J* = 0 and *K* = 1, then *J* = 0 when the *CK* input goes from 1 to 0. Finally, if both *J* and *K* are 1, then the output *toggles* or changes state just like a *T*-type flip-flop. Table 3–5 summarizes these operations, where the "1/0" notation signifies the logic 1 to 0 transition of the input clock.

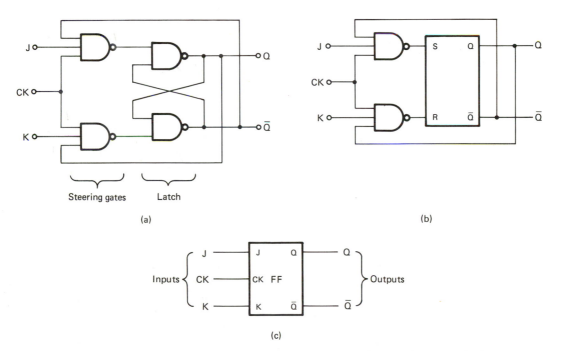

Fig. 3-15 *J–K* flip-flop: (a) logic diagram, (b) using a *R–S* flip-flop, (c) logic symbol.

Table 3-5
J–K Flip-Flop Truth Table

Input			Output		
J	K	CK	Q	$\overline{Q}$	Comments
0	0	1/0	No change		Latch
1	0	1/0	1	0	Set
0	1	1/0	0	1	Reset
1	1	1/0	Toggles		Outputs change

From the equivalent logic diagram of a *J–K* flip-flop, we should be aware that it is constructed using only NAND gates. As pointed out in Chapter 2, it takes a finite amount of time for a signal to travel through a logic gate, which we have called propagation delay. If, for example, it takes 5 nanoseconds (ns) for a signal to propagate through a NAND gate, then it takes 5 ns + 5 ns, or 10 ns, to propagate through the two gates comprising one-half of the flip-flop, that is, 5 ns for one of the steering gates, and 5 ns for one gate of the latch section. If both the *J* and *K* inputs are at logic 1, then this propagation delay of 10 ns could cause the flip-flop to toggle every 10 ns while a triggering signal is present at the clock input. Therefore, the clock pulses must be *shorter than 10 ns* or else the *J–K* flip-flop will repeatedly toggle with a frequency of 1/(10 ns), or 10 MHz, as long as there is a clock signal.

This situation where the flip-flop loses control is referred to as a *race condition*, or simply *racing*. Only when the input clock signal is removed will the flip-flop stop toggling back and forth. When it does stop, the output state is not predictable. Other than making sure that the input clock pulse is shorter than the propagation delay through the flip-flop, the best cure is to use an improvement, called the *master–slave J–K flip flop*.

2. Master–Slave *J–K* Flip-Flop

As shown in Figure 3–16, the master–slave *J–K* flip-flop is built using two flip-flop sections in series, with the clock input of the slave section being the complement of the clock input of the master section. In addition, the outputs of the slave flip-flop are connected in a cross-coupled feedback arrangement to the inputs of the master section flip-flop.

When the clock input of the master section goes from 0 to 1, the slave section is effectively disabled or frozen, since its clock input is 0. If

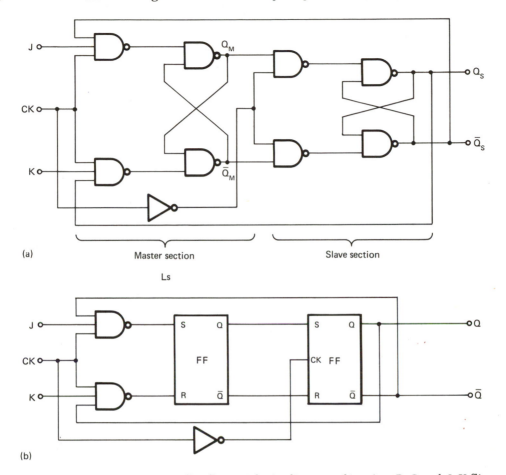

Fig. 3-16 Master–slave *J–K* flip-flop: (a) logic diagram, (b) using *R–S* and *J–K* flip-flops.

the slave section is disabled, any race condition that may exist in the master section now does not get passed onto the inputs of the slave section. The output of the slave drives or "steers" the master only while the clock is 1. If the slave is set ($Q_S=1$), the master can then only be reset ($Q_M=0$). On the other hand, if the slave is reset ($Q_S=0$), the master can only be set ($Q_M=1$). The truth table for the master–slave *J–K* flip-flop is the same as the basic edge-triggered *J–K* flip flop (Table 3–5).

How the master and slave sections interact can be summarized by the sequence diagram of Figure 3–17. The pulse shown is nonideal as it has finite rise and fall times. At point *A* on the rising or leading edge, the slave is disabled, so the master is isolated from the slave. At point *B*, the

Fig. 3-17 Master–slave flip-flop timing diagram.

action of the master flip-flop is controlled by the *J* and *K* inputs while the slave is still disabled. At point *C* on the falling or trailing edge, the master is disabled. Finally, at point *D*, the outputs of the master are transferred to the *J* and *K* inputs of the slave, which in turn determine the output states of the overall flip-flop.

Both the edge-triggered and master–slave *J–K* flip-flop can be further improved by providing either a *preset (PR)* input, a *clear (CLR)* input, or both. Alternate names for the preset input are *set* and *preset*, while *reset* and *preclear* have been occasionally used in place of clear. When either or both of these inputs are added, the *J*, *K*, and clock inputs are now more appropriately referred to as the *synchronous* inputs, since changes in the output of the flip-flop are synchronized with the 1 to 0 transition of the clock input and dependent on the logic levels of the *J* and *K* inputs.

On the other hand, both the preset and clear inputs are referred to as *asynchronous* inputs, since these inputs can independently override the collective actions of the *J, K,* and clock inputs. Figure 3–18 shows how both preset and clear inputs are added to an edge-triggered *J–K*

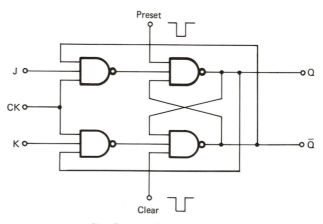

Fig. 3-18 *J–K* flip-flop with asynchronous preset and clear inputs.

flip-flop. As shown, the preset and clear inputs are connected directly to the latch section and consequently have the ability to override the synchronous inputs. If at any time the *preset* input is at logic 0 regardless of the signals at the *J, K,* and clock inputs, then output *Q* is immediately set at logic 1, which is the normal *set* state as far as the output is concerned. If output *Q* is already at logic 1 when the preset input is enabled, then *Q* stays at logic 1. Consequently, the preset input can be used to independently set or preset the flip-flop. On the other hand, if the *clear* input is at logic 0, output *Q* is then forced to a logic 0, which is the flip-flop's normal reset state. Therefore, the clear input can be used to independently reset or clear the flip-flop. Normally, *the preset and clear inputs are complementary* unless they are both at logic 1, in which case the flip-flop functions normally as a basic edge-triggered *J–K* flip-flop. The condition when the preset and clear inputs are both 0 should be avoided, as it is impossible for the flip-flop to be simultaneously set and reset. Figure 3–19 shows a master–slave *J–K* flip-flop with preset and clear inputs added to the latch portion of the slave section. As an example, the timing diagram of Figure 3–20 illustrates the interaction of the *J, K, CLK, PR,* and *CLR* inputs.

As summarized in Table 3–6, a wide selection of TTL edge-triggered and master–slave *J–K* flip-flops is available, some of which

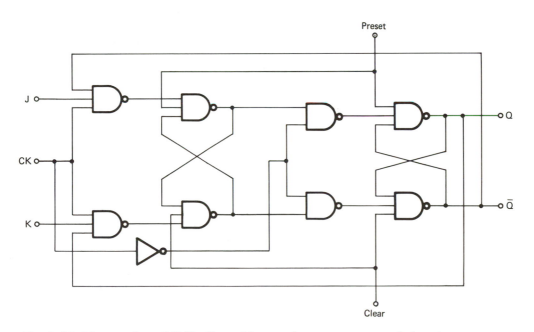

Fig. 3-19 Master–slave *J–K* flip-flop with asynchronous preset and clear inputs.

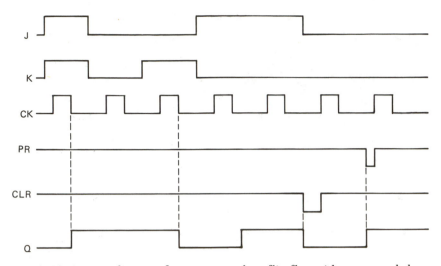

Fig. 3-20 Timing diagram for a master–slave flip-flop with preset and clear inputs.

have no preset and clear inputs, and some of which have preset only, clear only, or both. Table 3–7 lists the available CMOS *J–K* devices.

From Table 3–6, it should be noted that there are both positive and negative edge-triggered *J–K* flip-flops, as well as the master–slave version. As an example, Figure 3–21 shows a 7470 (positive edge-triggered), a 7473 (master–slave), and a 74106 (negative edge-triggered), all having their *J*, *K*, and *CK* inputs in parallel so that the operation of these three *J–K* types may be compared.

D-*Type Flip-Flop*

The *D*-type flip-flop (Figure 3–22a) has the same form as the *D*-type latch, except the gate input is now called the *clock* or *CK* input. If *D* is 1 when the clock signal goes from 0 to 1, then the flip-flop will be set, in which case *Q* = 1. As shown in the timing diagram of Figure 3–22c, the *transition* of the clock signal from 0 to 1 is termed the *positive edge*. Consequently, the flip-flop is triggered by the positive transition of the clock signal, which is the positive edge. On the other hand, if *D* = 0, the flip-flop is reset (*Q* = 0) on the positive edge of the clock signal. Therefore, the *D*-type flip-flop can be told to remember or store the state of the *D* input when a 0 to 1 transition is applied to the clock input.

A *D*-type flip-flop can be constructed from a *J–K* flip-flop by the addition of a single inverter, as shown in Figure 3–23. On the other

Table 3-6
Available TTL *J-K* Flip-Flops

Device	Function
7470	Positive edge-triggered with preset and clear
7471	Master–slave with preset
7472	Master–slave with preset and clear
7473	Dual master–slave with clear
7476	Dual master–slave with preset and clear
7478	Dual negative edge-triggered with preset, common clear, and clock
74101	Negative edge-triggered with preset
74103	Negative edge-triggered with clear
74106	Dual negative edge-triggered with preset and clear
74107	Dual master–slave with clear
74108	Dual negative edge-triggered with preset and clear
74109	Dual positive edge-triggered with preset and clear
74112	Dual negative edge-triggered with preset and clear
74113	Dual negative edge-triggered with preset
74114	Dual negative edge-triggered with preset and common clear and clock

Table 3-7
Available CMOS *J-K* Flip-Flops

Device	Function
74C73	Dual master–slave with clear
74C76	Dual negative-triggered with preset and clear
74C107	Dual master–slave with clear
4027	Dual master–slave with preset and clear

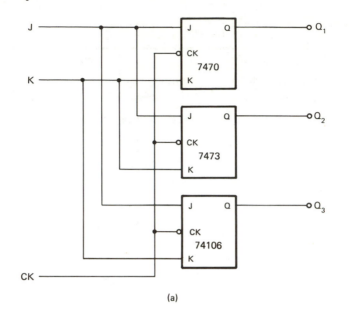

(a)

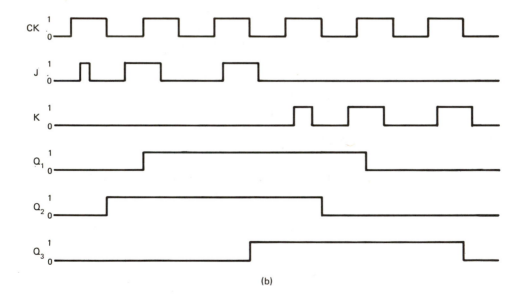

(b)

Fig. 3-21 Comparison of 7470, 7473, and 74106 flip-flops.

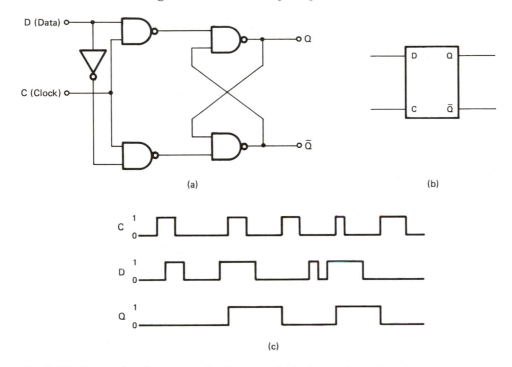

(a)   (b)

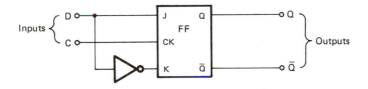

(c)

Fig. 3-22 *D*-type flip-flop: (a) logic diagram, (b) logic symbol, (c) timing diagram.

Fig. 3-23 *D*-type flip-flop using a *J–K* flip-flop: logic diagram.

hand, a *D*-type flip-flop can be made to toggle or function as a *T*-type flip-flop simply by connecting the $\overline{Q}$ output to the *D* input, as shown in Figure 3–24. Although we get one output pulse for every two input pulses, the action of the *D*-type is different than that of a *J–K* type connected as a toggle flip-flop, as shown in Figure 3–25. The *D*-type reacts to the positive transition of the clock pulse, whereas the *J–K* type reacts to the negative transition.

Table 3–8 summarizes the available TTL and CMOS *D*-type flip-flops. The 7474 is an exact pin-for-pin TTL replacement equivalent of

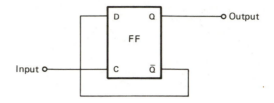

Fig. 3-24 Converting a *D*-type flip-flop to a *T*-type flip-flop.

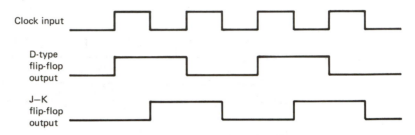

Fig. 3-25 Comparison of *D*-type and *J–K* flip-flops.

the 74C74, while the 4013 is a CMOS *functional* equivalent of the 7474. The 74174 and 74175 are exact pin-for-pin TTL replacements for the 74C174 and 74C175, while the 40174 and 40175 are the corresponding CMOS functional equivalents.

Table 3-8
TTL and CMOS *D*-Type Flip-Flops

Family	Device	Function
TTL	7474	Dual with preset and clear
TTL	74174	Hex with common clock and clear
TTL	74175	Quad with common clock and clear
CMOS	74C74	Dual with preset and clear
CMOS	40174/ 74C174	Hex with common clock and clear
CMOS	40175/ 74C175	Quad with common clock and clear
CMOS	4013	Dual with preset and clear

REVIEW

In this chapter we have discussed two basic elements of clocked logic, the latch and the flip-flop. Although they both serve as simple memory elements, their operations differ. In Chapter 5 we shall see how flip-flops are combined to create shift registers and counters, which permit the storage and transfer of digital data.

INTRODUCTION TO THE EXPERIMENTS

The following experiments are intended to illustrate the basic characteristics of several type of latches and flip-flops, using available integrated-circuit devices. The experiments that you will perform can be summarized as follows:

Experiment	Purpose
3–1	Demonstrates the operation of a 4-bit latch when connected to the output of a 7493 4-bit binary counter.
3–2	Compares the operation of the 7474 and 7475 devices as latches.
3–3	Demonstrates the operation of the 7476 *J–K* flip-flop with preset and clear.

EXPERIMENT 3–1

PURPOSE

The purpose of this experiment is to demonstrate the operation of a 4-bit *D*-type latch by storing the output of a 7493 4-bit binary counter.

PIN CONFIGURATION OF
INTEGRATED-CIRCUIT DEVICES

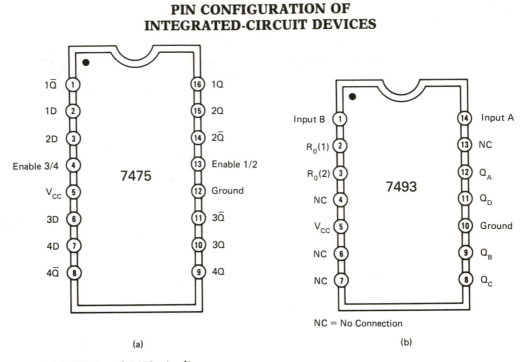

(a)   (b)

Fig. 3-26 7475 and 7493 pin diagrams.

Required Parts and Equipment

1 7475 TTL 4-bit latch
1 7493 TTL 4-bit (divide-by-16) counter
8 LED monitors
2 TTL logic pulsers
1 TTL square-wave clock (approximately 500 Hz)
1 solderless breadboard
1 TTL power supply

SCHEMATIC DIAGRAM OF CIRCUIT

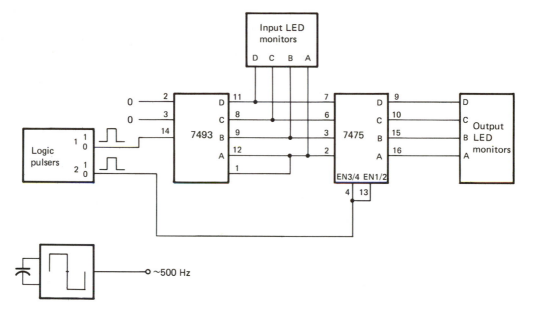

Fig. 3-27 Schematic diagram for Experiment 3-1.

STEP 1

Wire the circuit shown in the schematic diagram of Figure 3–27 and apply power to the breadboard.

STEP 2

First press in and release pulser 1 several times. You should notice that the binary code indicated by the LED monitors connected to the outputs of the 7493 counter increases upward. This is because the counter increases by one each time the pulser 1 is released. At the same time, what do you observe about the outputs of the 7475 4-bit latch?

As pulser 1 is activated several times, there should be *no change* in the output states of the 7475 latch. Since the *C* or clock (i.e, ENABLE) input to each of the internal four *D*-type flip-flops is 0, any data that are present at each flip-flop's *D* (data) input are ignored and do not get transferred to the output of that flip-flop.

STEP 3

Now press in pulser 2 so that the ENABLE inputs of the 7475 latch (pins 4 and 13) are at logic 1. At the same time, activate pulser 1 several times. You should notice that the corresponding inputs and outputs of the 4-bit latch are the same. As long as the *C* input of each *D*-type flip-flop is 1, the corresponding input and output states are the same.

STEP 4

Keeping pulser 2 pressed in, continue to activate pulser 1 until the four LED monitors indicate the 4-bit binary sequence *DCBA* = 0100, at which time release pulser 2. The corresponding input and output states of the latch should both have the binary sequence 0100.

Now continue to activate pulser 1 several more times. Do you observe any further change at the outputs of the 7475 latch?

You should have observed that the outputs of the 7475 latch indefinitely display the 4-bit sequence 0100 no matter what the 4-bit number is at the input of the latch. You have therefore *latched* or temporarily stored the 4-bit word 0100.

STEP 5

Keeping pulser 2 pressed in, again activate pulser 1 until the four LED monitors indicate the 4-bit binary sequence *DCBA* = 1010, at which time release pulser 2. The corresponding input and output states of the latch should both have the binary sequence 1010.

Now continue to activate pulser 1 several more times. Do you now observe any further change at the outputs of the 7475 latch?

You should have observed that the outputs of the 7475 latch indefinitely display the 4-bit sequence 1010 no matter what the 4-bit number is at the input of the latch. You have therefore latched or temporarily stored the 4-bit word 1010. Consequently, the *D*-type latch will store whatever is present at its input when the *C* or ENABLE input is brought back to logic 0 and will remain in this state until the ENABLE input is again at logic 1.

STEP 6

Instead of manually cycling the 7493 counter to produce any one of 16 possible 4-bit binary codes, remove pulser 1 and connect the output of a square-wave clock to the input (pin 14) of the 7493 counter. The square-wave frequency should be about 500 Hz. You should notice that all the LED monitors connected to the output of the counter appear to be lit, as the output states are changing at a rate that is faster than the eye can follow. If the input to the counter is 500 Hz, the *A* output of the counter (pin 12) is changing at a 250-Hz rate, while the *D* output (11) is changing at approximately a 31-Hz rate.

STEP 7

As in previous steps, press in pulser 2 and observe that the four LED monitors connected to the output of the latch appear to be lit. Now release pulser 2 and observe that you have now latched onto one of the 16 possible 4-bit binary numbers. As before, this action occurs at the instant that pulser 2 is released. This output now remains even though the inputs to the latch are still changing.

STEP 8

Repeat step 7 several times. You will probably display a different 4-bit number each time you do. It would be virtually impossible to predict which of the 16 possible 4-bit numbers will be latched when pulser 2 is released as the inputs change at a rate that is faster than the eye can follow. In Chapter 7 we will see how a series of 4-bit latches is used with digital displays, such as part of a frequency counter.

EXPERIMENT 3–2

PURPOSE

The purpose of this experiment is to simultaneously compare the operation of the 7474 *D*-type flip-flop, wired as a latch, and the 7475 latch.

PIN CONFIGURATION OF
INTEGRATED-CIRCUIT DEVICES

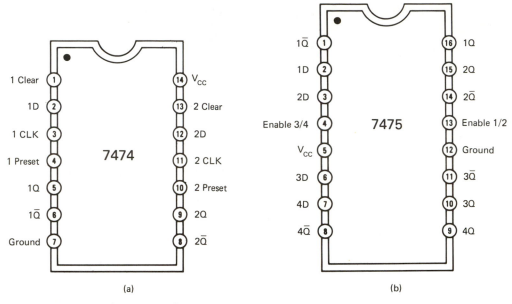

Fig. 3-28 7474 and 7475 pin diagrams.

Required Parts and Equipment

1 7474 quad *D*-type flip-flop
1 7475 TTL 4-bit latch
2 LED monitors
2 TTL logic pulsers
1 solderless breadboard
1 TTL power supply

SCHEMATIC DIAGRAM OF CIRCUIT

STEP 1

Wire the circuit shown in the circuit diagram of Figure 3–29 and apply power to the breadboard. Both LED monitors should be unlit, indicating that both outputs are at logic 0.

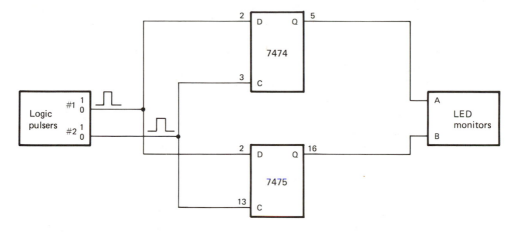

Fig. 3-29 Schematic diagram for Experiment 3-2.

STEP 2

Press the pulser that is connected simultaneously to the *D* or data inputs of each device (pulser 1). While pulser 1 is still pressed in, press and release *pulser 2*, which is connected to both *C* or clock inputs. In both cases, observe the state of the LED monitors. You should find that, in both cases, the outputs of the 7474 and 7475 are logic 1 when the *C* input goes from 0 to 1.

Now release the data pulser and then press the clock pulser once again. You should now see that both LED monitors are unlit. The timing diagram of Figure 3-30 illustrates the sequences taken in this step.

In this step, both devices behave in the same manner.

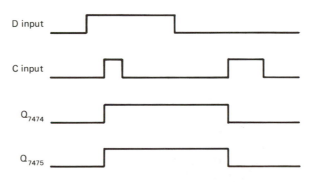

Fig. 3-30 Timing diagram for step 2.

STEP 3

Now, keeping the clock pulser pressed in, press and release the data pulser four times and observe what happens to the LED monitors. Press pulser 1 in for a fifth time and then release pulser 2. After pulser 2 is released, release pulser 1. Throughout this sequence, observe the LED monitors. You should have observed results that are summarized by the timing diagram in Figure 3–31. You should have noticed that the output of the 7474 flip-flop did not change during this sequence! Why?

The 7474 latches data on the *positive leading edge of the input clock pulse,* whereas the 7475 latches data on the *negative trailing edge* of the input clock pulse. In addition, the output of the 7474, while its clock input is at logic 1, *does not follow changes at its D input.* The output of the 7475 equals the state of its *D* input when the clock input is at logic 1.

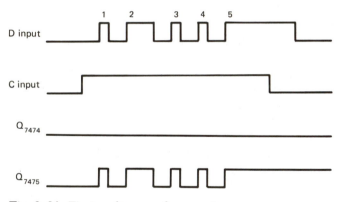

Fig. 3-31 Timing diagram for step 3.

EXPERIMENT 3-3

PURPOSE

The purpose of this experiment is to demonstrate the operation of a 7476 *J–K* flip-flop, observing the effect the settings of its synchronous and asynchronous inputs have on its *Q* output.

PIN CONFIGURATION OF INTEGRATED-CIRCUIT DEVICES

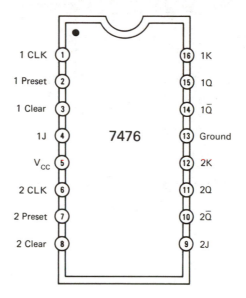

Fig. 3-32 7476 pin diagram.

Required Parts and Equipment

1   7476 TTL dual *J–K* flip-flop
1   LED monitor
3   TTL logic pulsers
2   TTL logic switches
1   solderless breadboard
1   TTL power supply

SCHEMATIC DIAGRAM OF CIRCUIT

STEP 1

Wire the circuit shown in the schematic diagram of Figure 3–33. Initially set both logic switches so that the *J* and *K* inputs are at logic 0. Pulser 1, which is connected to the flip-flop's CLOCK input, is normally at logic 0. The two pulsers that are connected to the PRESET and CLEAR inputs are normally at logic 1.

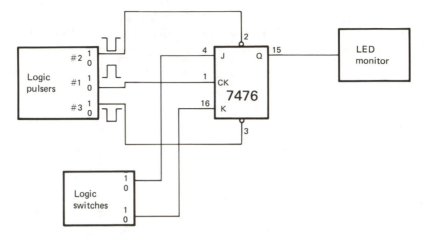

Fig. 3-33 Schematic diagram for Experiment 3-3.

STEP 2

Apply power to the breadboard. While watching the LED monitor, repeatedly press and release pulser 1, which is the clock input signal to the flip-flop. What happens to the *Q* output?

You should observe that, with both *J* and *K* inputs at logic 0, repeated clocking of the *J–K* flip-flop has no effect on the output states.

STEP 3

Now set the logic switches so that *J* = 1 and *K* = 0. Press and release pulser 1 several times. What now happens to the *Q* output?

You should notice that the *Q* output changes from 0 to 1 the first time pulser 1 is released. Repeated activation of the pulser has no further effect on the *Q* output. Therefore, if *J* = 1 and *K* = 0, then *Q* = 1 on the 1-to-0 transition of the input clock pulse, which is frequently referred to as the *negative* or trailing edge of the clock pulse.

STEP 4

Now set the logic switches so that *J* = 0 and *K* = 1. Press and release the pulser several times. What now happens to the *Q* output?

You should notice that the Q output changes from 1 to 0 the first time pulser 1 is released. Repeated activation of the pulser has no further effect on the Q output. Therefore, if $J = 0$ and $K = 1$, then $Q = 0$ on the negative edge of the clock pulse.

STEP 5

Now set the logic switches so that both the J and K inputs are at logic 1. Press and release pulser 1 ten times. What now happens to the Q output?

You should notice that when the pulser is released the Q output *toggles* or changes state. You should also notice that the output sequence repeats itself every two times the clock pulser is released. That is, when the Q output goes from 0 to 1, you have to press and release the pulser two more times before the output again goes from 0 to 1.

By having both J and K inputs at logic 1, we now have a T-type or *toggle* flip-flop. Another name for a T-type flip-flop is a *binary divider* or *divide-by-2 counter,* as the output frequency is always one-half that of the input. Consequently, if the clock pulser is repeatedly pressed and released ten times, the Q output will cycle between 0 and 1 five times.

The J and K inputs are sometimes referred to as *synchronous* inputs. Depending on the setting of the J and K inputs, the output can change state on the negative transition of the input clock pulse.

STEP 6

Press and release the pulser so that the Q output is 0. Now press and release the pulser connected to the 7476's PRESET input (pin 2). Continue to press and release this pulser. What happens to the Q output?

When the PRESET input is briefly at logic 0, the flip-flop is then *set*, so $Q = 1$. This action takes place independent of the J, K, and CLOCK inputs. Since the Q output is already at logic 1, continued pressing and releasing of the pulser has no further effect on the Q output. The PRESET input is also sometimes referred to as the SET input.

STEP 7

Now, with $Q=1$, press and release the pulser connected the 7476's CLEAR input (pin 3). Continue to press and release this pulser. What happens to the Q output?

When the CLEAR input is briefly at logic 0, the flip-flop is then cleared, or *reset*, so $Q=0$. This action also takes place independent of the J, K, and CLOCK inputs. Since the Q output is now already at logic 0, continued pressing and releasing of the pulser has no further effect on the Q output.

The PRESET and CLEAR inputs are usually referred to as *asynchronous* inputs, as either action is independent of the input clock pulse. These inputs are used to initialize the output to 0 or 1 prior to any action by the input clock. Based on your observations, now complete the truth table for the 7476 *J–K* flip-flop.

7476 *J–K* Flip-Flop Truth Table

Input					Output
PRESET	CLEAR	*J*	*K*	CLOCK	*Q*
1	1	0	0	⎍	
1	1	1	0	⎍	
1	1	0	1	⎍	
1	1	1	1	⎍	
0	1	×	×	×	
1	0	×	×	×	

× = don't care.

MULTIVIBRATORS

OBJECTIVES

After completing this chapter, you will be able to understand:

- *The function of monostable and astable multivibrators.*

- *How monostable and astable multivibrators are built using discrete logic elements.*

- *How to use TTL and CMOS integrated circuits to function as either a monostable or astable multivibrator.*

- *How to build crystal-controlled astable multivibrators.*

MONOSTABLE MULTIVIBRATORS

The monostable multivibrator, which is frequently referred to as a *one-shot*, is a device whose output has *only one stable state*. When it is triggered, the one-shot's output immediately changes from its stable state (e.g., logic 0) to its unstable state (logic 1) for a given amount of time and then returns back to the stable state. In this section we will

103

study how to construct monostable multivibrators using discrete logic gates, as well as look at those one-shots that are conveniently available as a single integrated circuit.

Discrete Monostables

Figure 4–1 shows a discrete monostable multivibrator made using a 2-input NOR gate, an inverter, a resistor, and a capacitor. Both the NOR gate and inverter may either be TTL or CMOS, but not a combination of TTL and CMOS devices. For a trigger pulse briefly going from 0 (the *stable* state) to 1 and back to 0, called a *positive trigger pulse*, the output of the NOR gate goes *low* or 0. Coupling through the capacitor–resistor combination makes the output of the inverter go *high* (the unstable state), which is then fed back to the remaining input of the NOR gate and keeps the NOR gate's output low. It should be noted that we could easily have used another 2-input NOR gate with both inputs tied together in place of the simple inverter.

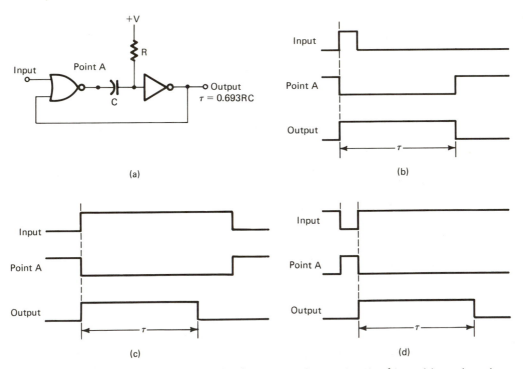

Fig. 4-1 NOR gate monostable multivibrator: (a) basic circuit, (b) positive triggering, (c) positive trigger pulse width longer than output pulse width, (d) negative triggering.

After this series of events, the timing capacitor C begins to charge exponentially toward the positive logic supply voltage through the timing resistor R. When the voltage reaches a level that is sufficient to appear equal to a logic 1 to the inverter, the output of the inverter returns to 0 (the stable state). How fast or slow this process is carried out depends on the *RC time constant*. When this NOR gate monostable circuit is triggered with a positive pulse, the output will remain high (logic 1) for a given amount of time, called the *time delay or output pulse width*, τ, which is basically given by the equation

$$\tau = 0.693RC, \text{ seconds (s)} \qquad (4\text{–}1)$$

So if $R = 100 \text{ k}\Omega$ and $C = 0.47 \text{ }\mu\text{F}$, the output pulse width is

$$\tau = 0.693(100 \text{ k}\Omega)(0.47 \text{ }\mu\text{F})$$
$$= 0.0326 \text{ s}$$
$$= 32.6 \text{ ms}$$

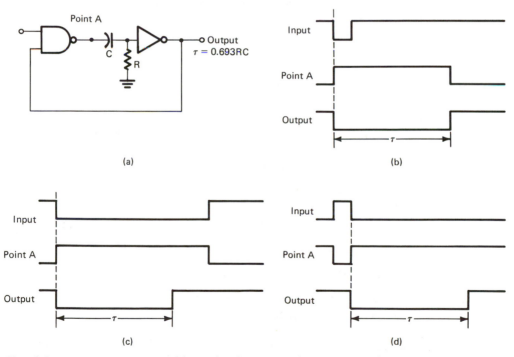

Fig. 4-2 NAND gate monostable multivibrator: (a) basic circuit, (b) negative triggering, (c) negative trigger pulse width longer than output pulse width, (d) positive triggering.

As illustrated in the timing design of Figure 4–1b, this circuit responds to the *positive-going leading edge* of the trigger pulse. Equation 4–1 holds true only when the input trigger pulse is *longer* than the desired output pulse width. On the other hand, if the trigger width is *shorter* than the output, the output of the NOR is just the complement of the input trigger, and no time delay occurs.

Figure 4–1d illustrates the timing diagram for a *negative* trigger pulse. Note that the timing period does not start until the *positive-going trailing edge* of the trigger pulse occurs, although the stable state of the output pulse is still logic 0.

Figure 4–2 illustrates a *negatively triggered* monostable multivibrator using a 2-input NAND gate. It should be noted that the timing resistor is now tried to ground (logic 0). In addition, the output's stable state is logic 1, and goes low for a time period given by Eq. 4–1 when triggered.

As a practical matter, one-shots made from discrete logic elements are not widely used, particularly when the output pulse width is to be accurately repeated. In addition, these simple discrete monostables are not *retriggerable*. That is, they do not have the capability to be repeatedly triggered and thereby extend the output pulse width, even though the output has not yet returned to its stable state.

Integrated-Circuit Monostables

Instead of constructing discrete monstable circuits where the output pulse might be unpredictable, it is more advantageous to consider using one of a number of available *monolithic* integrated-circuit monostable devices as an easier, but superior, alternative.

1. TTL Devices

The following TTL monostable multivibrators have found widespread use:

- 74121 Monostable multivibrator
- 74122 Retriggerable monostable multivibrator with clear
- 74123 Dual retriggerable monostable multivibrator with clear

These three devices are best used for short and accurate pulse widths, ranging from nanoseconds up to milliseconds, and when complementary outputs are required.

• 74121 Monostable

As shown by the pin diagram in Figure 4–3, the 74121 has three inputs (*A*1, *A*2, and *B*), two complementary outputs, and three *timing* pins. Figure 4–4a shows how the 74121 is connected for *positive* triggering. The output pulse width, as a function of the timing resistor and capacitor, is the same as Eq. 4–1:

$$\tau = 0.693RC \quad \text{seconds}$$

For optimum performance, *R* should be in the 2- to 40-kΩ range, while *C* should be between 10 picofarads (pF) and 10 μF. Figure 4–4b shows a graph of possible output pulse widths as a function of various timing resistor and capacitor values. For example, suppose we want the 74121 wired to produce an output pulse width of approximately 20 μs. Using the graph, we find that several resistor–capacitor combinations are possible. The two most obvious ones are

- 3 kΩ and 0.01 μF

- 30 kΩ and 0.001 μF (i.e., 1000 pF)

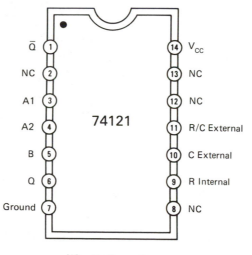

NC = No Connection

Fig. 4-3 74121 TTL monostable multivibrator pin diagram.

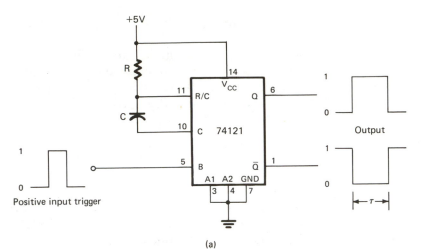

(a)

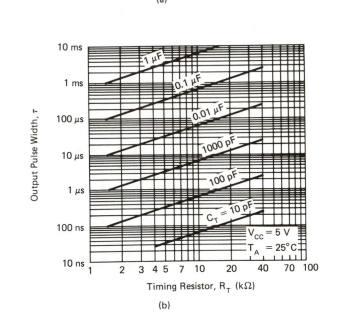

(b)

Fig. 4-4 74121 monostable multivibrator connected for positive triggering: (a) basic circuit, (b) graph of output pulse width.

which can be verified by using Eq. 4–1:

$$\tau = 0.0693RC$$
$$= 0.693(3 \text{ k}\Omega)(0.01 \ \mu\text{F})$$
$$= 20.8 \ \mu\text{s}$$

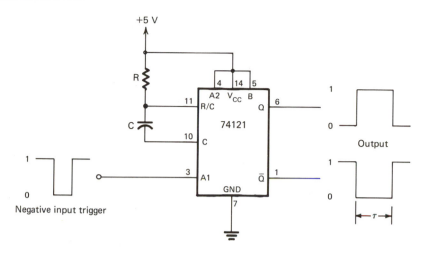

Fig. 4-5 74121 monostable multivibrator connected for negative triggering.

Alternatively, instead of using the graph, the required values for the timing components may be determined by using Eq. 4–1. Since there is not as wide a selection for capacitor values as there is for resistors, it is best to first select a standard *capacitor* value and then calculate the appropriate timing resistor value for a given pulse width. For example, to have a 4.5-ms pulse width using a 0.22-μF capacitor, we would need a resistance of 29.5 kΩ.

Figure 4–5 shows how to connect the 74121 to accept a *negative* input trigger pulse. As before, the output pulse width is determined either by the graph of Figure 4–4b or by Eq. 4–1.

• **74122 Monostable**

The 74122, whose pin diagram is shown in Figure 4–6, is an improvement over the 74121. The two major differences between them are that the 74122 has a CLEAR input and that it is retriggerable.

The CLEAR input is used to reset the monostable. Normally, the CLEAR input is at logic 1. When brought *low* (logic 0), the multivibrator is prevented or inhibited from triggering. If the 74122 has been already triggered, the complementary outputs immediately return or are reset to their stable states. Consequently, the *Q* output returns to 0.

The circuit shown in Figure 4–7a shows how the 74122 is connected for *positive* triggering. The output pulse width, as a function of the timing resistor and capacitor, is given by

$$\tau = 0.32RC\left[1 + \frac{700}{R}\right] \text{ seconds} \qquad (4\text{-}2)$$

For optimum performance, R should be in the 5- to 50-kΩ range, while C should be at least 10 pF. Figure 4–7b shows a graph of possible output pulse widths as a function of various timing resistor and capacitor values.

For example, suppose we want the 74122 wired to produce an output pulse width of approximately 2 μs. From the graph of Figure 4–7b, we find that the combination of 20 kΩ and 300 pF will do the job, although other combinations are possible. On the other hand, if we first select a suitable capacitor value, such as 470 pF, then the required timing resistance is

$$R = \frac{\tau}{0.32C} - 700 \ \Omega$$

$$= 13,000 - 700$$

$$= 12.6 \text{ k}\Omega$$

The timing diagram of Figure 4–7c shows the action of the CLEAR input on the Q output. Figure 4–7d shows the action of an additional positive trigger pulse applied *before* the output pulse width is

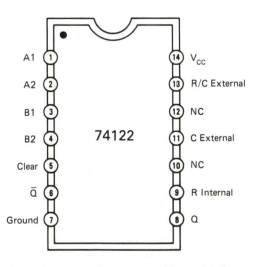

Fig. 4-6 74122 TTL monostable multivibrator pin diagram.

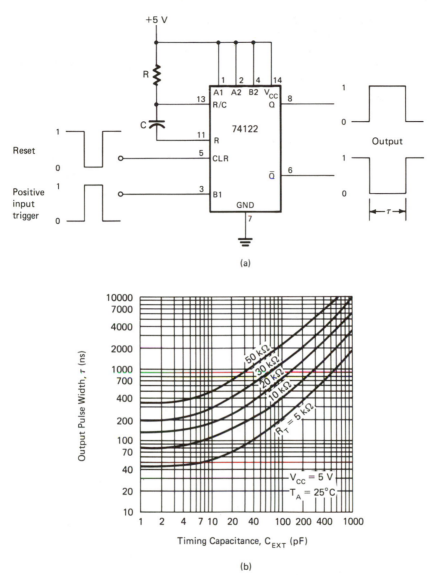

(a)

(b)

Fig. 4-7 74122 monostable multivibrator connected for positive triggering: (a) basic circuit, (b) graph of output pulse width, (c) timing diagram, (d) effect of retriggering.

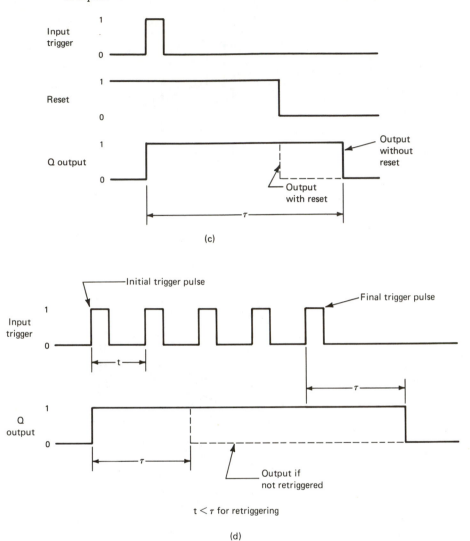

(c)

(d)

$t < \tau$ for retriggering

Fig. 4-7 (*Continued*)

terminated. This is called *retriggering* and can be used to extend the output pulse width.

For negative triggering, the required connections for the 74122 are shown in Figure 4–8. As before, the CLEAR input is normally kept at +5 V, unless we want to reset the one-shot.

Internally, the 74122 contains the equivalent of a 10-kΩ resistor whose external connection is the R_{INT} terminal (pin 9). This *internal*

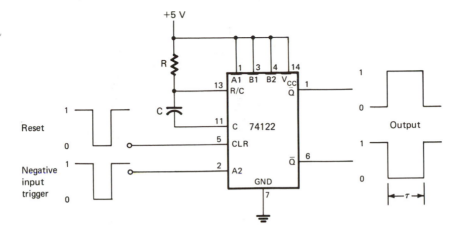

Fig. 4-8 74122 monostable multivibrator connected for negative triggering.

resistor can be used in place of an external resistor, as shown in Figure 4–9, by connecting pin 9 to +5 V, and can be used for both positive and negative triggering modes. The external timing capacitor is connected between pins 11 and 13.

• 74123 Monostable

The remaining TTL one-shot is the 74123, whose pin diagram is shown in Figure 4–10; it is basically nothing more than the equivalent of

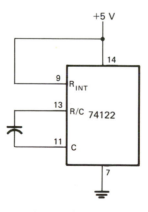

Fig. 4-9 74122 monostable multivibrator using the internal 10-kΩ resistor.

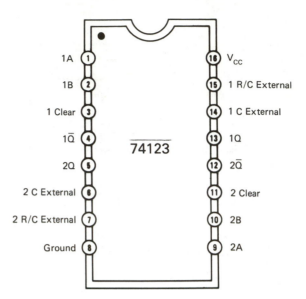

Fig. 4-10 74123 TTL dual monostable multi-vibrator pin diagram.

two 74122 one-shots in a single package. Both monostable sections have CLEAR inputs and are capable of being retriggered. The connections required for positive triggering of a single section are shown in Figure 4–11a, and the connections required for negative triggering are shown in Figure 4–11b. In either case, the CLEAR input is normally high.

The nominal output pulse width, as a function of the timing resistor and capacitor, is given by

$$\tau = 0.28RC\left[1 + \frac{700}{R}\right] \text{ seconds} \qquad (4\text{–}3)$$

For optimum performance, R should be in the 5- to 50-kΩ range, while C should be at least 10 pF.

2. CMOS Devices

The following CMOS monostable multivibrators are presently available:

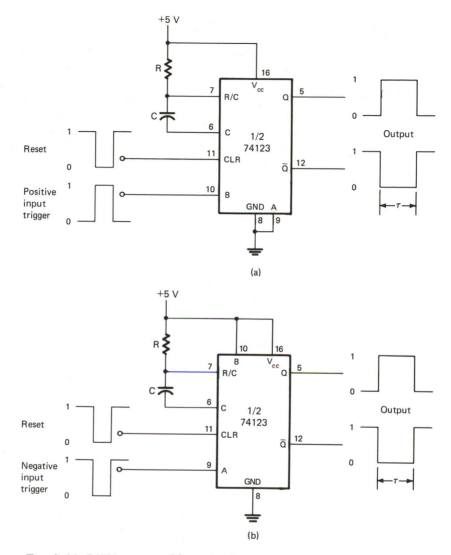

(a)

(b)

Fig. 4-11 74123 monostable multivibrator circuits: (a) positive triggering, (b) negative triggering.

- 4047 Monostable/astable multivibrator
- 4098/MC14528 Dual retriggerable and resettable monostable multivibrator
- 74C221 Dual retriggerable monostable multivibrator with clear

These are discussed individually.

• 4047 Monostable

The 4047, whose pin diagram is shown in Figure 4–12, possesses a high degree of accuracy and stability when compared with discrete monostable circuits. Like the TTL monostables, the 4047 also provides complementary outputs. Positive edge triggering is accomplished using the circuit of Figure 4–13a, while the negative edge trigger circuit is shown in Figure 4–13b. In either case, the minimum input trigger pulse is typically 500 ns for a 4-V supply, 200 ns for a 10-V supply, and 140 ns for 15 V.

When triggered, the output pulse width is given by

$$\tau = 2.48RC \quad \text{seconds} \qquad (4\text{--}4)$$

However, for reliable operation, the timing resistor should be in the 10-kΩ to 1-MΩ range, while the capacitor can be any convenient value greater than 0.001 μF. Consequently, the output pulse width can be as small as 25 μs.

• 4098/MC14528 Monostable

As shown in the pin diagram of Figure 4–14, the 4098 (or Motorola's equivalent MC14528) is a dual monostable in a single package and is functionally identical to the 74123 TTL device. Each

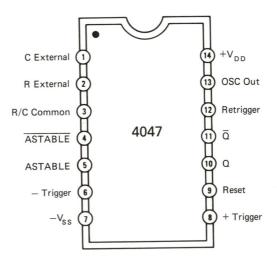

Fig. 4-12 4047 CMOS dual monostable multivibrator pin diagram.

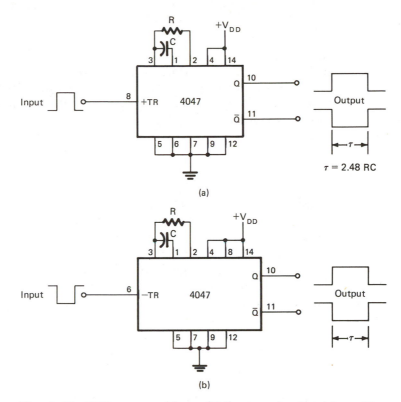

(a)

(b)

Fig. 4-13 4047 monostable multivibrator circuits: (a) positive triggering, (b) negative triggering.

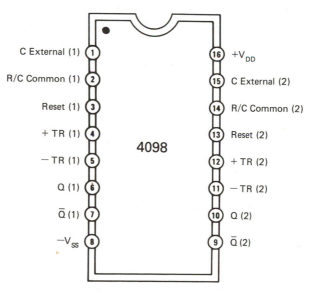

Fig. 4-14 4098 CMOS dual monostable multivibrator pin diagram.

section has both a positive leading edge ($+TR$) and a negative trailing edge ($-TR$) input in addition to complementary outputs.

Figure 4–15 shows the connections required for positive edge *nonretriggering* of either section. Once triggered, the output pulse width is given by

$$\tau = 0.5RC \quad \text{seconds} \tag{4–5}$$

The timing capacitor C should be in the 0.01- to 100-μF range, while the minimum value for R is 5 kΩ. Since this is a *nonretriggerable* one-shot configuration, the timing cycle will go to completion as if triggered by a single pulse, even if the $+TR$ input is repeatedly triggered during the timing cycle. If a negative pulse is applied to the RESET input, the timing cycle is immediately terminated; otherwise, the RESET input should be connected to $+V_{DD}$ if not used. On the other hand, if an entire monostable section is not used in a particular application, the RESET and $+TR$ inputs are connected to $-V_{SS}$ (ground), while the $-TR$ input must be connected to $+V_{DD}$ to prevent excessive device power dissipation and possible destruction.

Figure 4–16 shows the required connections for positive edge retriggering operation. When connected in this mode, Eq. 4–5 determines the output pulse width *after* the last trigger edge is encountered.

The connections required for the corresponding negative trailing edge triggering circuits are shown in Figure 4–17. Taking into account the extreme values allowed for the external timing components, Table 4–1 summarizes the typical allowable input positive or negative trigger and reset pulse widths for the 4098/MC14528.

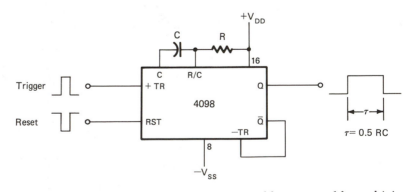
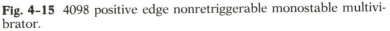

Fig. 4-15 4098 positive edge nonretriggerable monostable multivibrator.

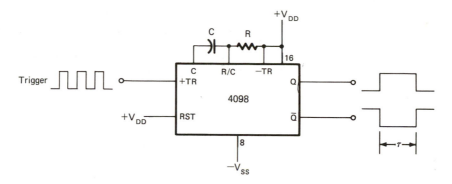

Fig. 4-16 4098 positive edge retriggerable monostable multivibrator.

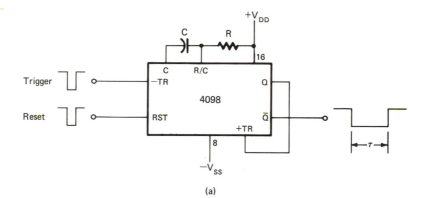

(a)

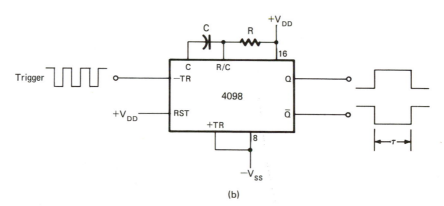

(b)

Fig. 4-17 4098 negative edge triggerable monostable multivibrator: (a) nonretriggerable, (b) retriggerable.

Table 4-1
Typical Minimum Trigger and Reset Pulse Widths for the 4098/MC14528 Monostable

Supply Voltage, V	Trigger, ns	Reset, μs
5	70	25
10	30	15
15	20	10

• 74C221 Monostable

The 74C221, unlike the 4098/MC14528 one-shot, is an exact pin-for-pin CMOS equivalent of the 74123 TTL dual monostable (Figure 4–10). However, its output pulse width is not the same as Eq. 4–3, but is given by

$$\tau = 0.994RC \quad \text{seconds} \tag{4–6}$$

which is frequently approximated by $\tau = RC$ for most situations.

ASTABLE MULTIVIBRATORS

The output of an astable multivibrator has two states, but neither is stable. Its output therefore alternates back and forth between 0 and 1, producing a constant stream or train of pulses, sometimes referred to as *clock pulses*, as shown in Figure 4–18. Since these clock pulses are continually being generated at a given rate, the astable multivibrator is often thought of as a *free-running* pulse generator.

The *period (T)* of the output pulses is the time taken to complete one cycle, that is, the time it takes when the waveform starts its 0-to-1 transition (point *A*, Figure 4–18) to the next time the clock pulse makes a 0-to-1 transition (point *C*). The *output frequency (f)* of these clock pulses is found from

$$f = \frac{1}{t} \quad \text{hertz (Hz)} \tag{4–7}$$

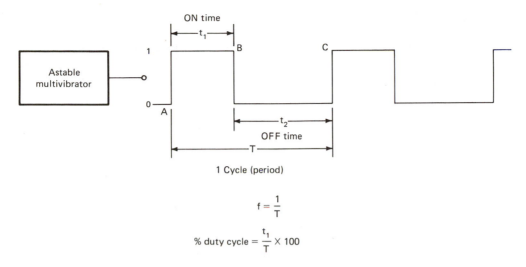

Fig. 4-18   Astable multivibrator clock pulses showing duty cycle.

Consequently, for example, if the period is 250 μs, the astable frequency is then

$$f = \frac{1}{250 \ \mu s}$$

$$= 4000 \ \text{Hz} \quad \text{or} \quad 4 \ \text{kHz}$$

The time segment $A\text{–}B\,(t_1)$ of a given cycle is referred to as the ON time, while segment $B\text{–}C\,(t_2)$ is the OFF time. A measure of how long the clock pulse remains in a given state is called the *duty cycle*. In digital logic, the duty cycle can take on one of several definitions. Perhaps the most often used is to take the *ratio of the ON time to the time needed to complete the cycle*; this is generally expressed as a percentage. Thus

$$\% \ \text{Duty cycle} = \frac{t_1}{T} \times 100 \qquad (4\text{–}8)$$

Consequently, the duty cycle is independent of frequency. If the ON and OFF times are exactly equal, then the duty cycle is 50%, in which case the clock pulse train is *symmetrical*. In this special case the pulse train is called a *square wave*.

Discrete Astables

Figure 4–19 shows a simple astable multivibrator made using two inverters with a single *RC* time constant; it is useful for frequencies up to about 10 MHz. The output frequency is given by the approximate relation

$$f \approx \frac{0.721}{RC} \ \text{Hz} \qquad\qquad (4\text{--}9)$$

The circuit is somewhat stable, can be made with either TTL or CMOS devices, has an approximate 50% duty cycle, and will oscillate by itself when power is first applied. The inverter function may be replaced by NAND or NOR equivalent functions.

One drawback with the circuit of Figure 4–19 if CMOS inverters (or their equivalent function) are used is its frequency sensitivity to supply voltage changes. The widely used CMOS astable circuit of Figure 4–20 minimizes this sensitivity by the addition of a second resistor R_s, which is made from two to ten times the value of R. The output frequency is found from

$$f = \frac{0.455}{RC} \ \text{Hz} \qquad\qquad (4\text{--}10)$$

The timing resistor R is restricted to the 4-kΩ to 1-MΩ range, while C can be any value greater than 100 pF for reliable operation.

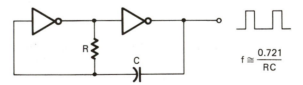

Fig. 4–19 Basic astable multivibrator circuit using inverters.

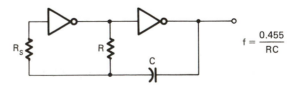

Fig. 4–20 Improved astable multivibrator.

Certain applications require that the clock pulses from an astable multivibrator have a required symmetry. Using CMOS inverters, the circuit of Figure 4–20 can be modified to include potentiometer control of the duty cycle, as shown in Figure 4–21. The resistance of the potentiometer is the sum of the fractional resistance R_A and R_B. If $R_A + R_B$ is made large compared to the forward resistance of the diode, typically up to several hundred ohms, the percent duty cycle is then given by

$$\% \text{ Duty cycle} = \frac{R_A}{R_A + R_B} \times 100 \qquad (4\text{--}11)$$

If the resistance R_A is less than R_B, the duty cycle will be less than 50%; if greater than R_B, it will be greater than 50%.

The output frequency for such a circuit is

$$f = \frac{1.443}{(R_A + R_B)C} \quad \text{Hz} \qquad (4\text{--}12)$$

assuming that $R_A + R_B$ is very large compared to the forward resistance of two diodes in series. Like the circuit of Figure 4–20, resistor R_S is made from two to ten times $R_A + R_B$ to minimize power supply variations on the output frequency.

Other discrete astable circuits are possible based on variations of the basic circuit of Figure 4–19. However, these do not offer any significant advantage or improvement over those integrated-circuit devices specifically designed to function as astable multivibrators.

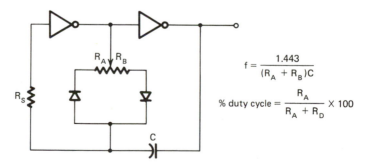

$$f = \frac{1.443}{(R_A + R_B)C}$$

$$\% \text{ duty cycle} = \frac{R_A}{R_A + R_D} \times 100$$

Fig. 4-21 Fixed-frequency astable multivibrator with duty cycle control.

Integrated-Circuit Astables

Like monostables, a superior alternative to discrete astable circuits is to use one of several integrated-circuit devices designed for this purpose. In general, no 7400 series TTL devices are available. However, the general practice when designing TTL circuits is to use an *integrated circuit timer*, such as the *555 timer* described later in this chapter.

For CMOS circuits, the 4047 (Figure 4–12) discussed previously also functions as an astable multivibrator by having a logic 1 at both the ASTABLE (pin 5) and $\overline{\text{ASTABLE}}$ inputs (pin 4), as shown in Figure 4–22. For this circuit, there are two output frequencies. When the output is taken from the OSC OUTPUT (pin 13), the nominal output frequency is determined from the equation

$$f = \frac{1}{2.2RC} \quad \text{Hz} \tag{4–13}$$

However, this output wave form will not necessarily have a 50% duty cycle. On the other hand, *complementary symmetrical square waves* are obtained at pins 10 and 11, but they are *one-half* that of the OSC OUTPUT, or

$$f = \frac{1}{4.4RC} \quad \text{Hz} \tag{4–14}$$

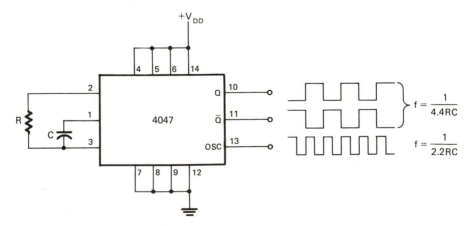

Fig. 4-22 4047 astable multivibrator having three outputs.

For oscillator frequencies up to about 500 kHz at pin 13, the external timing capacitor can be any value greater than 100 pF, while the timing resistor should be in the 10-kΩ to 1-MΩ range.

The 4098/MC14528 CMOS dual monostable device (Figure 4–14) can be wired so that one independent monostable section triggers the remaining section, as shown in Figure 4–23. The ON time (t_1) is controlled by R_1 and C_1 (Eq. 4–5), while the OFF time (t_2) is set by R_2 and C_2; thus

$$t_1 = 0.5 R_1 C_1 \quad \text{seconds}$$

$$t_2 = 0.5 R_2 C_2 \quad \text{seconds}$$

Consequently, the output frequency is

$$f = \frac{1}{t_1 + t_2}$$

$$= \frac{2}{(R_1 C_1 + R_2 C_2)} \quad \text{Hz} \qquad (4\text{–}15)$$

CRYSTAL-CONTROLLED ASTABLES

The crystal-controlled astable, often called simply a crystal oscillator, allows the generation of a more stable and accurate train of clock

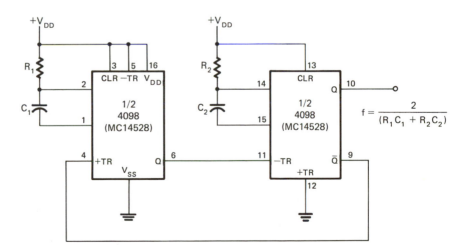

Fig. 4-23 4098 astable multivibrator with independent ON and OFF times.

pulses. Since the output frequency is set by the crystal, changes in frequency due to timing components, supply voltage, and temperature variations are minimized. In general, crystal-controlled astables are used for crystal frequencies in the 1- to 10-MHz range, but they can produce lower frequencies when used with divide-by-N counters, as discussed in Chapter 5.

Figure 4–24a shows a widely used crystal-controlled TTL astable, built using three sections of a 7404 hex inverter. The variable capacitor is used to *trim* or adjust the output frequency so that it is equal to the crystal frequency. Such an adjustment can be used to vary the frequency of oscillation by several percent of the crystal frequency. A variation of this basic TTL circuit using NAND gates is shown in Figure 4–24b. In both circuits, inverter $G3$ is used as a buffer to isolate the oscillator from the possible loading effects of additional circuitry that is connected to its output.

A crystal-controlled CMOS oscillator, using a pair of inverters, is shown in Figure 4–25. Inverter $G1$ is the oscillator section, while the

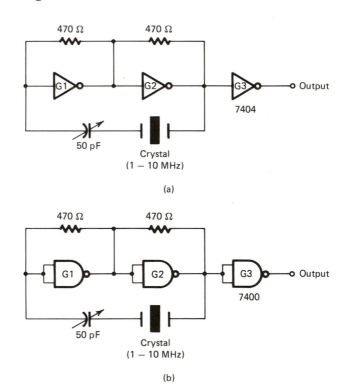

Fig. 4-24 Crystal-controlled TTL astable multivibrator: (a) using inverters, (b) using NAND gates.

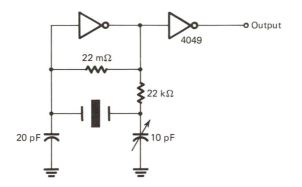

Fig. 4-25   Crystal-controlled CMOS astable multi-vibrator.

second inverter serves as a buffer. The variable capacitor, like the circuit of Figure 4–24, is used to set the output frequency equal to the crystal frequency. Other CMOS devices, such as the 4055 and 4060, can be used to create crystal-controlled oscillators. Since these devices have several possible output frequencies, controlled by divide-by-N counters, they are discussed in Chapter 5.

IC TIMERS

Since their introduction, a class of devices called *integrated circuit timers* has made the design of monostable and astable multivibrator circuits a simple task.[*] Of the myriad of available IC timers, perhaps the most popular is the 555 timer and its dual version, the 556 timer, whose pin diagrams are shown in Figure 4–26. As it is capable of operating with supply voltages from 4 to 18 V, it can be used with either TTL or CMOS devices. Although the standard 555 timer may have a current drain that may make it unsuitable for portable, battery-powered applications, low-power CMOS versions (L555, ICM7555, etc.) that are pin-for-pin equivalents of the standard timer are available.

---

[*]These low-cost devices have proved to be popular, and several books have been written describing their operation and application. See H. M. Berlin, *The 555 Timer Applications Sourcebook, with Experiments,* Howard W. Sams & Co. Inc., Indianapolis, 1976; J. J. Carr, *The IC Timer Handbook,* Tab Books, Blue Ridge Summit, Pa., 1981; W. G. Jung, *IC Timer Cookbook,* Howard W. Sams & Co., Inc., Indianapolis, 1977.

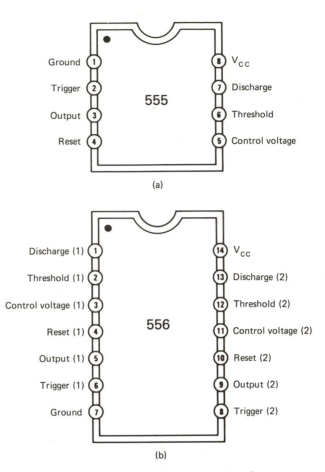

Fig. 4-26 Integrated-circuit timer pin diagrams: (a) 555 timer, (b) 556 dual timer.

Figure 4-27 shows the basic *negatively triggered* 555 timer monostable circuit. Based on the selection of a single external timing resistor and capacitor, the output at pin 3, when triggered, will be at logic 1 for a time period given by

$$\tau = 1.1RC \quad \text{seconds} \qquad (4\text{--}16)$$

For reliable operation, the timing resistor should be in the 1-kΩ to 10-MΩ range, while the timing capacitor can be any practical value greater than 0.001 μF.

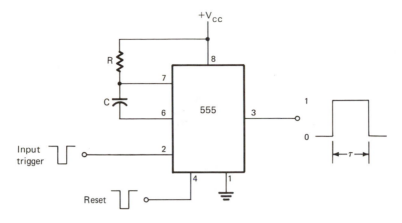

Fig. 4-27 Negatively triggered 555 monostable multivibrator.

Once triggered, the output will remain at logic 1 until the set time has elapsed, even if retriggered during the timing cycle. If, on the other hand, a logic 0 is applied to the reset input (pin 4) during the timing cycle, the timing cycle is halted and the output is reset to logic 0.

Most integrated one-shots require a certain amount of time to recover after triggering. If this recovery time is not completed, the next timing cycle may be less than that given by Eq. 4–16. To eliminate this possibility, the negative recovery circuit of Figure 4–28 should be used. As long as the trigger pulse train keeps arriving at a given rate less than the 555's set time delay, the one-shot stays triggered and the output stays at logic 1. Once retriggering stops, the timing cycle goes to completion.

As shown in Figure 4–29, the 555 timer also functions as an astable multivibrator. The output frequency is given by

$$f = \frac{1.443}{(R_1 + 2R_2)C} \quad \text{Hz} \qquad (4\text{--}17)$$

The output pulse train will have a percent duty cycle that depends solely on the values of the two external timing resistors, so that

$$\% \text{ Duty cycle} = \frac{R_1 + R_2}{R_1 + 2R_2} \times 100 \qquad (4\text{--}18)$$

Thus, by making R_2 very large with respect to R_1, the duty cycle approaches 50%. The resistance sum $R_1 + 2R_2$ should be in the 1-kΩ to 10-MΩ range. The maximum practical output frequency is about 200

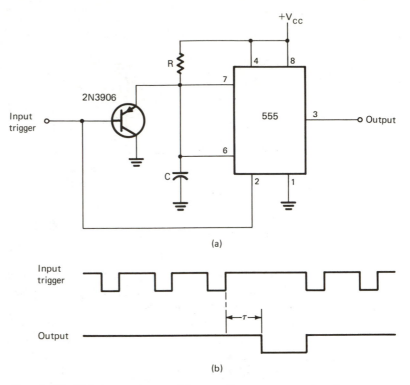

(a)

(b)

Fig. 4-28 555 timer monostable negative recovery circuit.

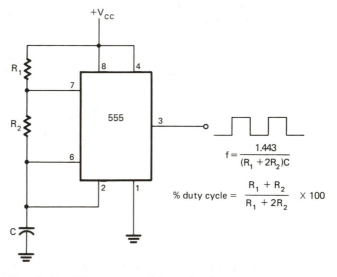

$$f = \frac{1.443}{(R_1 + 2R_2)C}$$

$$\% \text{ duty cycle} = \frac{R_1 + R_2}{R_1 + 2R_2} \times 100$$

Fig. 4-29 555 timer astable multivibrator.

kHz. It should be emphasized that the output duty cycle cannot be independently adjusted without changing the output frequency.

REVIEW

This chapter discussed a number of monostable and astable multivibrator circuits for both TTL and CMOS applications. Although it is possible to construct multivibrators using discrete logic gates, resistors, and capacitors, it more advantageous to create multivibrators using one of a number of devices designed specifically for this purpose, rather than reinvent the wheel. On the other hand, IC timers, such as the 555 timer, offer more flexibility than most TTL and CMOS multivibrators.

INTRODUCTION TO THE EXPERIMENTS

The following experiments are designed to demonstrate the operation of several devices that can be wired as monostable or astable multivibrators. The experiments that you will perform can be summarized as follows:

Experiment	Purpose
4–1	Demonstrates the operation of the 4047 CMOS multivibrator connected as a monostable multivibrator, triggered either by positive- or negative-going pulses.
4–2	Demonstrates the operation of the 555 timer as a monostable multivibrator.
4–3	Demonstrates the operation of the 4047 wired as an astable multivibrator.
4–4	Demonstrates the operation of the 555 timer as an astable multivibrator.

EXPERIMENT 4–1

PURPOSE

The purpose of this experiment is to demonstrate the operation of the 4047 CMOS multivibrator connected as a monostable multivibrator, triggered either by positive- or negative-going pulses.

PIN CONFIGURATION OF
THE 4047 MULTIVIBRATOR

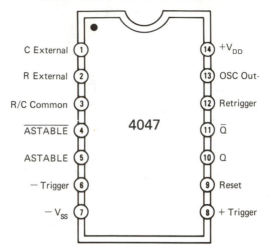

Fig. 4-30 4047 pin diagram.

Required Parts and Equipment

1 220-kΩ resistor
1 330-kΩ resistor
1 10-μF capacitor
1 4047 CMOS multivibrator

2 digital pulsers
2 LED monitors
1 solderless breadboard
1 power supply (9 V)

1 stopwatch or wristwatch with a second hand

SCHEMATIC DIAGRAM OF CIRCUIT

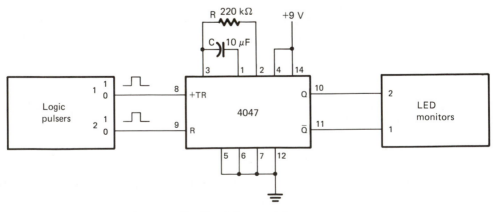

Fig. 4-31 Schematic diagram for Experiment 4-1.

DESIGN BASICS

$$\text{Output pulse width:} \quad \tau = 2.48RC$$

STEP 1

Wire the circuit as shown in the schematic diagram, using a 220-kΩ resistor for R and a 10-μF electrolytic capacitor for C. Then apply power to the breadboard.

STEP 2

Using either a stopwatch, digital wristwatch, or a watch with a second hand, determine how many seconds LED monitor 2 stays lit when pulser 1 is quickly pressed and released.

Initially, you should observe that LED monitor 1 is lit and LED monitor 2 is unlit. When pulse 1 is pressed and released, LED monitor 1 immediately goes out, while LED monitor 2 lights up. The multivibrator is now *triggered.* You should have determined that the LED monitor stays lit for aproximately 6 s.

This circuit is triggered by *positive* trigger pulses, as it responds to the logic 0 to logic 1 transition of the trigger pulse. For this reason, the input trigger is connected to the +TR input (pin 8) of the 4047.

STEP 3

Disconnect the power from the breadboard. Replace the 220-kΩ timing resistor with a 390-kΩ resistor, and then reconnect the power to the breadboard. Again, quickly press and release pulser 1 and determine the time that LED monitor 2 is lit.

You should have measured approximately 10 s. Confirm this with the equation given under *Design Basics.*

STEP 4

Again, quickly press and release pulser 1. Then press and release pulser 2, which is connected to the 4047's reset input. What happens to the LED monitors?

When pulser 2 is pressed, LED monitor 2 goes out, while LED monitor 1 lights up. When the reset input changes from logic 0 to logic 1, the timing cycle is then terminated, so the Q output (LED 2) returns to logic 0 (unlit), while the $\overline{Q}$ output returns to logic 1. At any time during the timing cycle, the cycle can be halted by setting the reset input at logic 1.

STEP 5

Disconnect the power from the breadboard. Now connect pin 8 to the positive supply voltage, and connect pin 6 (the −TR input) to the output of pulser 1, which is normally at logic 1. This circuit will now be triggered by *negative* trigger pulses.

Apply power to the breadboard. Now quickly press and release pulser 1 and measure the time that LED monitor 2 is lit.

You should find that LED monitor 2 remains lit for approximately 10 s, as was determined in step 3.

STEP 6

Repeat step 4. Is the result the same as was observed in step 4?

The two results are the same, since the 4047 monostable is reset when the reset input is at logic 1, regardless of positive or negative triggering.

EXPERIMENT 4-2

PURPOSE

The purpose of this experiment is to demonstrate the operation of the 555 timer as a monostable multivibrator.

PIN CONFIGURATION OF 555 TIMER

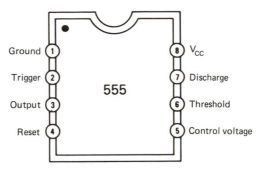

Fig. 4-32 555 timer pin diagram.

Required Parts and Equipment

1 330-kΩ resistor
1 1-MΩ resistor
1 10-μF capacitor
1 555 IC timer
2 digital pulsers
1 LED monitor
1 solderless breadboard
1 power supply (9 V)
1 stopwatch or wristwatch with a second hand

SCHEMATIC DIAGRAM OF CIRCUIT

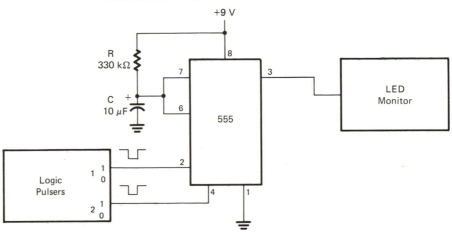

Fig. 4-33 Schematic diagram for Experiment 4-2.

DESIGN BASICS

Output pulse width: $\tau = 1.1\,RC$

STEP 1

Wire the circuit shown in the schematic, first using a 330-kΩ resistor for R and a 10-μF electrolytic capacitor for C. Apply power to the breadboard.

STEP 2

Using either a stopwatch, digital wristwatch, or a watch with a second hand, determine how many seconds the LED monitor stays lit when the pulser is quickly pressed and released.

You should have determined that the LED monitor stays lit for approximately 3.6 s.

STEP 3

Now press the pulser for approximately 5 to 10 s while watching the LED monitor.

You should observe that the LED monitor remained lit until the pulser was released, or perhaps the LED monitor remained lit approximately 3.6 s after you released the pulser. The reason is that the time duration of the trigger pulse exceeded the timer's time delay. Consequently, the negative trigger pulse should be kept short compared with the desired output pulse width.

STEP 4

Disconnect the power from the breadboard. Replace the timing resistor with a 1-MΩ resistor and apply power to the breadboard. Repeat step 2.

The LED monitor should be lit for about 11 s. Again, quickly press and release the pulser. Then quickly press the pulser that is connected to the reset input (pin 4) of the 555 timer.

You should observe that the LED monitor is immediately unlit. From this action, it should be clear that the reset function inhibits the timer. Consequently, the 555 timer can be reset at any time during the timing cycle simply by grounding its reset input. To trigger the timer again, pin 4 must be at logic 1.

EXPERIMENT 4-3

PURPOSE

The purpose of this experiment is to demonstrate the operation of the 4047 wired as an astable multivibrator.

PIN CONFIGURATION OF THE 4047 MULTIVIBRATOR

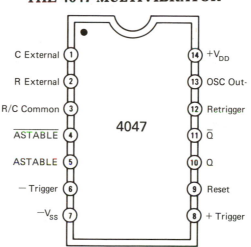

Fig. 4-34 4047 pin diagram.

Required Parts and Equipment

1   10-kΩ resistor
1   1-MΩ resistor
1   0.01-μF capacitor
1   0.33-μF capacitor
1   4047 CMOS multivibrator
3   LED monitors
1   solderless breadboard
1   power supply (5 V)
1   oscilloscope or frequency counter
1   stopwatch or wristwatch with a second hand

SCHEMATIC DIAGRAM OF CIRCUIT

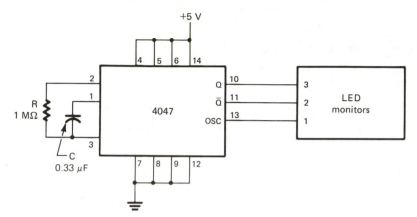

Fig. 4-35 Schematic diagram for Experiment 4-3.

DESIGN BASICS

Output frequency: $f = \dfrac{1}{2.2\,RC}$ (pin 13)

$$f = \dfrac{1}{4.4\,RC}$$ (pin 10 or 11)

STEP 1

Wire the circuit shown in the schematic diagram and apply power to the breadboard.

STEP 2

Using either a stopwatch, digital wristwatch, or a watch with a second hand, count the number of times LED monitor 1 flashes on and off in a 1-min period.

Depending on the accuracy of your timing components, you should count about 82 flashes in a 1-min period, which is the same as a frequency of 1.37 Hz and is slow enough for the eye to follow.

STEP 3

Now count the number of times LED monitors 2 and 3 each flash on and off over a 1-min period. How does this value compare with what you determined in step 2?

You should observe that LED monitors 2 and 3 both flash at a rate that is *one-half* the rate you measured in step 2.

Do you notice any relationship between LED monitors 2 and 3?

You should observe that, when LED monitor 2 is off, LED monitor 3 is on, so they are complements of each other.

STEP 4

Now disconnect the power from the breadboard and replace the 0.33-μF capacitor with a 0.01-μF capacitor. Also replace the 1-MΩ resistor with a 10-kΩ resistor. Reconnect the power to the breadboard and connect either an oscilloscope or a frequency counter to pin 13 of the 4047. Measure the output frequency. How does it compare with the equation given under *Design Basics?*

Depending on the accuracy of your timing components, the output frequency at pin 13 should be approximately 4545 Hz.

Now measure the frequencies at pins 10 and 11. How do they compare with the frequency at pin 13?

The frequency at pins 10 and 11 should be exactly one-half that of pin 13, or approximately 2273 Hz. Your reading will probably differ by 1 Hz since the frequency counter has an error of 1 Hz.

EXPERIMENT 4-4

PURPOSE

The purpose of this experiment is to demonstrate the operation of the 555 timer as an astable multivibrator.

PIN CONFIGURATION OF THE 555 TIMER

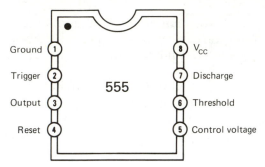

Fig. 4-36 555 timer pin diagram.

Required Parts and Equipment

1   1-kΩ resistor

1   100-kΩ resistor

1   330-kΩ resistor

1   1-MΩ resistor

1   0.001-μF capacitor

1   0.01-μF capacitor

1   1-μF capacitor

1   555 IC timer

1   solderless breadboard

1   power supply (5 V)

1   oscilloscope

1   stopwatch or wristwatch with a second hand

SCHEMATIC DIAGRAM OF CIRCUIT

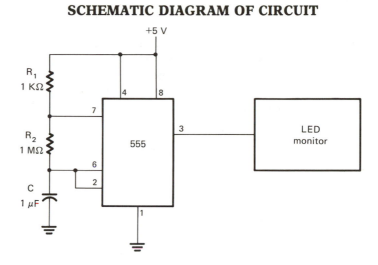

Fig. 4–37 Schematic diagram for Experiment 4-4.

DESIGN BASICS

$$\text{Output frequency:} \quad f = \frac{1.433}{(R_1 + 2R_2)C}$$

$$\text{Percent duty cycle:} \quad \%D = \frac{R_1 + R_2}{R_1 + 2R_2} \times 100$$

STEP 1

Wire the circuit shown in the schematic diagram.

STEP 2

Connect the power to the breadboard and observe that the LED monitor periodically flashes on and off. Using either a stopwatch, digital wristwatch, or watch with a second hand, count the number of times the LED monitor flashes on and off over a 1-min period.

You should count about 44 flashes per minute, which is the same as a frequency of 0.73 Hz and is slow enough for the eye to follow.

STEP 3

Disconnect the power from the breadboard. Change the resistor R_2 to 330 kΩ and the timing capacitor to 0.001 μF, and connect a frequency counter to the timer's output pin (pin 3). Apply power to the breadboard. Now measure the output frequency and compare the measured frequency with the equation under *Design Basics*.

You should measure an output frequency of approximately 4373 Hz.

STEP 4

Again disconnect the power from the breadboard. Now use a 100-kΩ resistor for R_2 and a 0.01-μF timing capacitor. In addition, connect an oscilloscope to pin 3 of the 555 timer.

STEP 5

Apply power to the breadboard. What is the output frequency and the duty cycle of the output pulse train?

You should measure a frequency of approximately 720 Hz. The duty cycle is the ratio of the time period that the output is at logic 1 (+5 V) to the time for one complete cycle. In this case, you should find that the duty cycle is very nearly 50%. Compare these results with the equations given under *Design Basics*.

5

SHIFT REGISTERS AND COUNTERS

INTRODUCTION

The temporary storage and transfer of data is an important function of most digital systems, especially computers. Shift registers are used primarily for storing and shifting entered data. Counters, like shift registers, also store data, but they also have the ability to tell how many input pulses they have received. Without counters, the vast array of digital instruments and appliances, such as time clocks, CB radios, digital voltmeters, and frequency counters, could not be built. Without the shift register, the parallel-to-serial transfer of data between computer systems could not be accomplished, nor could their computational operations.

OBJECTIVES

After completing this chapter, you will understand:

- *How shift registers differ from flip-flops.*
- *The differences between SISO, SIPO, PISO, and PIPO shift registers.*
- *The operation of a universal shift register.*
- *The operation of binary ripple counters.*

- *The operation of synchronous binary counters.*
- *The operation of a decade counter.*
- *How to cascade counters to produce a given modulus.*
- *The operation of programmable counters.*
- *The operation of shift register counters.*
- *How to multiply and divide frequencies using rate multipliers.*
- *How to generate waveforms using digital counters.*

SHIFT REGISTERS

A *shift register* is a synchronous clocked system using cascaded flip-flops that permit the *storage and transfer* of digital data. Since shift registers are classified as serial or parallel, we shall discuss them separately.

Serial in–Serial out Shift Register

A *serial in–serial out* or *SISO* shift register accepts or processes data in a *serial* format. This means that data are entered 1 bit at a time, and data are output 1 bit at a time.

A *4-bit* SISO shift register is shown in Figure 5–1; it uses four cascaded *D*-type flip-flops, each having a clear input, such as the 74175, 74C175, and 40175 devices. Since either a *R–S* or a *J–K* flip-flop can be wired as a *D*-type, the SISO shift register can also be constructed using these types. Since there are four flip-flops, the *storage capacity* of this arrangement is 4 bits, hence its name, the 4-bit register. There is a single

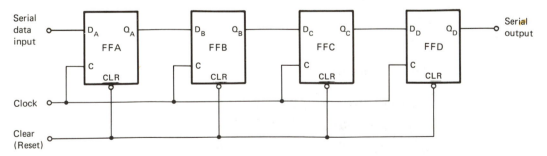

Fig. 5-1 Four-bit SISO shift register using *D*-type flip-flops.

data, clear, and clock input, while the two outputs are complementary. To explain how such a SISO shift register works, let's look at how the 4-bit binary number 1001 is processed.

Initially, let's briefly set the clear input at logic 0 so that each of the four flip-flops is *reset;* that is, outputs $Q_D = Q_C = Q_B = Q_A = 0$. Starting with the rightmost or least significant bit (1) at the data input, flip-flop *A* (*FFA*) will then be *set* when the first applied clock pulse goes from 0 to 1. Consequently, $Q_A = 1$, so the 1 is now *stored* by *FFA* as indicated by the state of Q_A. The second bit (0) is then applied at the data input. When the second clock pulse is applied, *FFB* is now *set,* since $Q_B = D_B = 1$. Consequently, the data bit (i.e., 1) held by *FFA* is now *shifted right* from *FFA* to *FFB*, as $Q_B = 1$. At the same time, since *FFA* is now reset, $D_A = 0$, so $Q_A = 0$. Therefore, the second data bit is now shifted into *FFA*. At this point, $Q_B = 1$ and $Q_A = 0$.

The third bit (0) is applied to the data input. Since $D_C = 0$, $D_B = 0$, and $D_A = 1$, *FFC*, *FFB*, and *FFA*, respectively, will be in a reset–reset–set sequence when the third clock pulse is applied. This corresponds to an output sequence of $Q_C = 1$, $Q_B = 0$, and $Q_A = 0$ so that the first bit has now been shifted three times through the register. Finally, when the last data bit (1) is applied to the data input and a clock pulse applied, *FFA* will then be *set,* while *FFB*, *FFC*, and *FFD* will now be in a reset–reset–set sequence. Consequently, $Q_D = 1$, $Q_C = 0$, $Q_B = 0$, and $Q_A = 1$, which is the original 4-bit number input to the shift register. The fourth clock pulse then completes the serial entry of the 4-bit number 1001, where it is then stored indefinitely as long as power is supplied. The shifting of these 4 bits serially into the register is illustrated in Figure 5–2.

To retrieve these 4 bits from the shift register, the bits must be shifted out serially. The easiest way to accomplish this is to set the data input at logic 0 and apply three clock pulses. Initially, output Q_D will be 1, followed in turn by 0, 0, and 1. On the fourth clock pulse, the output will be 0 as the input data bit (a logic 0) has now been shifted through four flip-flops. Since the data input is at logic 0 throughout this entire process, the net result is that the shift register is now cleared (i.e., all outputs at 0) after the 4-bit number has been serially shifted out of the register, as shown in Figure 5–3.

Figure 5–4 shows the formation of a 4-bit SISO shift register using *J–K* flip-flops. In this case, *FFA* is wired as a *D*-type flip-flop, while the complementary outputs of *FFB* and *FFC* are connected to the *J* and *K* inputs of *FFC* and *FFD*, respectively.

Instead of physically connecting the four sections of the 74175 quad *D*-type flip-flop together to create a 4-bit SISO register, we can use one of several devices that are designed for this purpose, such as the

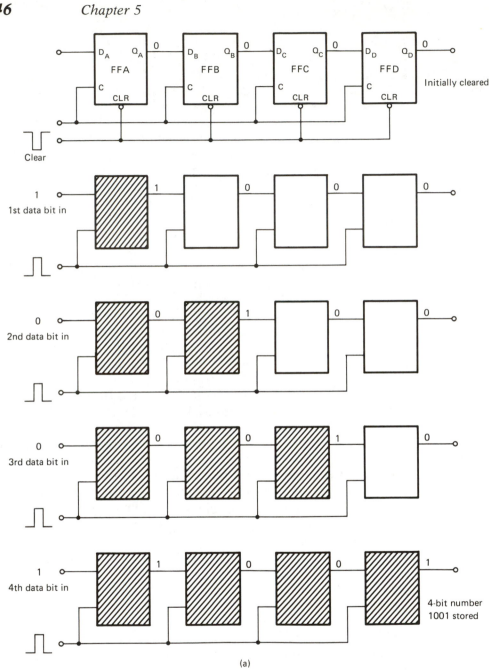

(a)

Fig. 5-2 Four bits serially entered into a SISO shift register: (a) flip-flop shifting, (b) timing diagram.

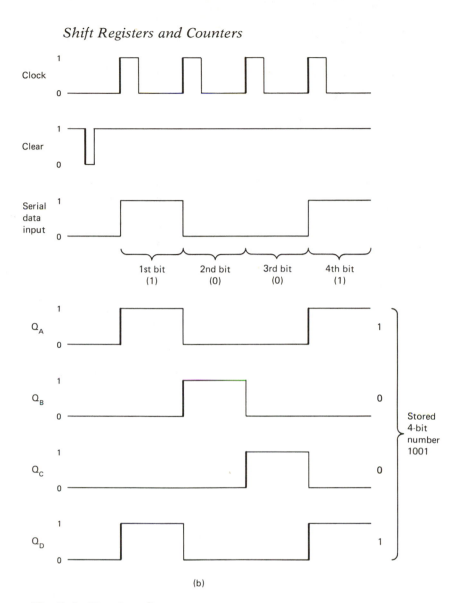

(b)

Fig. 5-2 *(Continued)*

7494 TTL device. As shown by the internal logic diagram in Figure 5–5b, the 7494 is a little more versatile than the basic arrangement of Figure 5–1. In addition to the data (*D*), clock (*CK*), and clear (*CLR*) inputs, the 7494 also has two preset enable inputs (*PE*1 and *PE*2), as well as four preset input pairs (*P*1*A* and *P*1*B*, *P*2*A* and *P*2*B*, etc.).

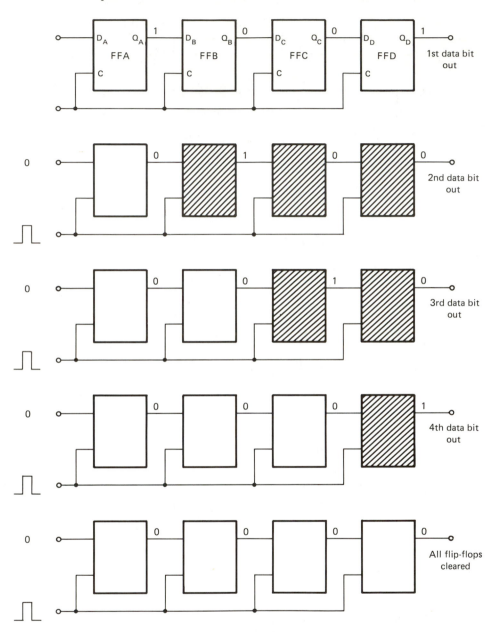

Fig. 5-3 Four bits serially shifted out of a SISO shift register.

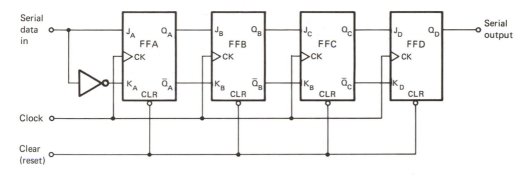

Fig. 5-4 Four-bit SISO shift register using *J–K* flip-flops.

All flip-flops are simultaneously reset so that $Q = 0$ when the *CLR* input is briefly set at logic 1 while the preset inputs are all 1. The 7494 may then be parallel loaded, or *preset*, to a given 4-bit binary number by first using the *CLR* input followed by the setting one of the preset input pairs to this number. Clearing all four stages sets the output of each stage to 0. Data to be loaded are applied to either the *P1A, P2A, P3A, P4A* input group or the *P1B, P2B, P3B, P4B* input group, with the corresponding preset enable input at 1. The action of clearing and presetting the register is *asynchronous* or independent of the clock pulse. However, once cleared or preset, *CLR* must be 0 and the preset inputs at 1 for normal SISO shift register operation to occur with each clock pulse. Since the 7494 has only a *preset* capability, so that a logic 1 of any given stage cannot be changed to 0 using the preset function. The register must be first cleared so that the outputs of *all* stages are 0, and then the given stages are preset to a logic 1.

As with flip-flops, shift register packages like the 7491 8-bit SISO shift register can be cascaded to form higher-capacity registers. Shown in Figure 5–6 are two 8-bit SISO register packages cascaded to form a 16-bit SISO shift register. The serial output of the first register package goes to the serial input of the second package.

Serial in–Parallel out Shift Register

A *serial in–parallel out* or *SIPO* shift register has the same flip-flop arrangement as a SISO register, except that the output of each flip-flop is available, as shown in Figure 5–7, using *D*-type flip-flops. Therefore, as each bit is shifted and stored, the output of each flip-flop indicates the respective data. Rather than outputting the data on a serial basis, the SIPO register outputs all the data bits simultaneously, hence the

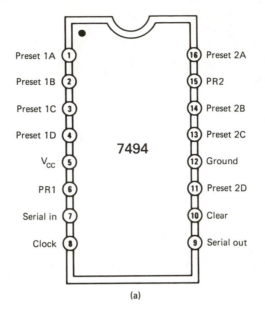

(a)

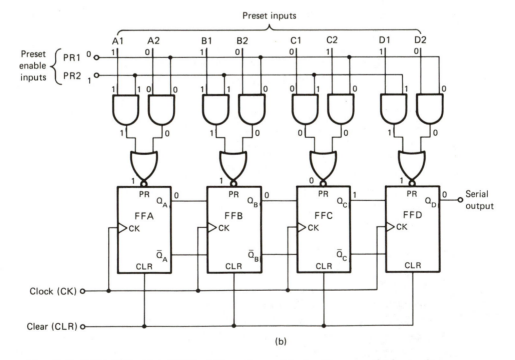

(b)

Fig. 5-5 7494 TTL 4-bit SISO shift register: (a) pin diagram, (b) logic diagram.

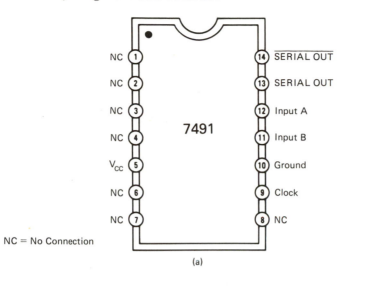

NC = No Connection

(a)

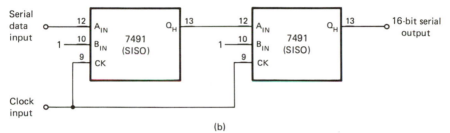

(b)

Fig. 5-6 7491 TTL 8-bit SISO shift register: (a) pin diagram, (b) two 7491s cascaded for a 16-bit SISO shift register.

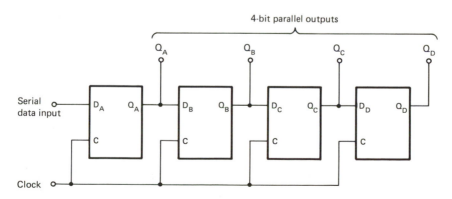

Fig. 5-7 Four-bit SIPO shift register.

term, *parallel* output. Another way of looking at a SIPO register is as a serial-to-parallel converter, which finds widespread applications in digital computers.

As an example, the 74164 is a TTL 8-bit SIPO shift register (equivalent to the CMOS 74C164), whose pin and logic diagrams are shown in Figure 5–8. Although all flip-flops are of the *R–S* type, FFA nevertheless is still connected as a *D*-type. The 74164 has two serial inputs NANDed together so that one input can be used to gate the other. Using the 8-bit binary number 10010011, the operation of the 74164 is illustrated by the timing diagram of Figure 5–9. Like the SISO register, any number of SIPO registers may be cascaded to increase the overall capacity of a shift register system.

Parallel in–Serial out Shift Register

Instead of serially entering and shifting data 1 bit at a time through a series of cascaded flip-flops, a *parallel in–serial out* or *PISO* shift register accepts all the input data bits simultaneously, or in *parallel* form. Once the multibit data are stored, they are output serially, so we have, in effect, a parallel-to-serial converter.

The logic diagram for a typical 4-bit PISO shift register is shown in Figure 5–10. All four *D*-type flip-flops are connected together by a common clock input. Data present at the four inputs D_A, D_B, D_C, and D_D are stored by their respective flip-flops when the clock pulse is applied, and the parallel enable (*PE*) input is at logic 1, which enables the AND gates connected to these data inputs. To output data serially, the *PE* input is first brought to logic 0, which (1) disables the parallel input AND gates, and (2) enables the AND gates connected from the outputs of *FFA*, *FFB*, and *FFC*. Three clock pulses are then required to shift the remaining 3 bits through the register. A timing diagram describing the parallel loading and serial outputting of the 4-bit binary number 1001 is shown in Figure 5–11.

Parallel in–Parallel out Shift Register

The final basic register type is the *parallel in–parallel out* or *PIPO* shift register. As shown in Figure 5–12, this 4-bit PIPO shift register, in its simplest form, is basically nothing more than four *D*-type latches that have a common clock input. Consequently, the 4 data bits that are simultaneously present at inputs D_A, D_B, D_C, and D_D immediately appear on the parallel outputs when all flip-flops are simultaneously clocked.

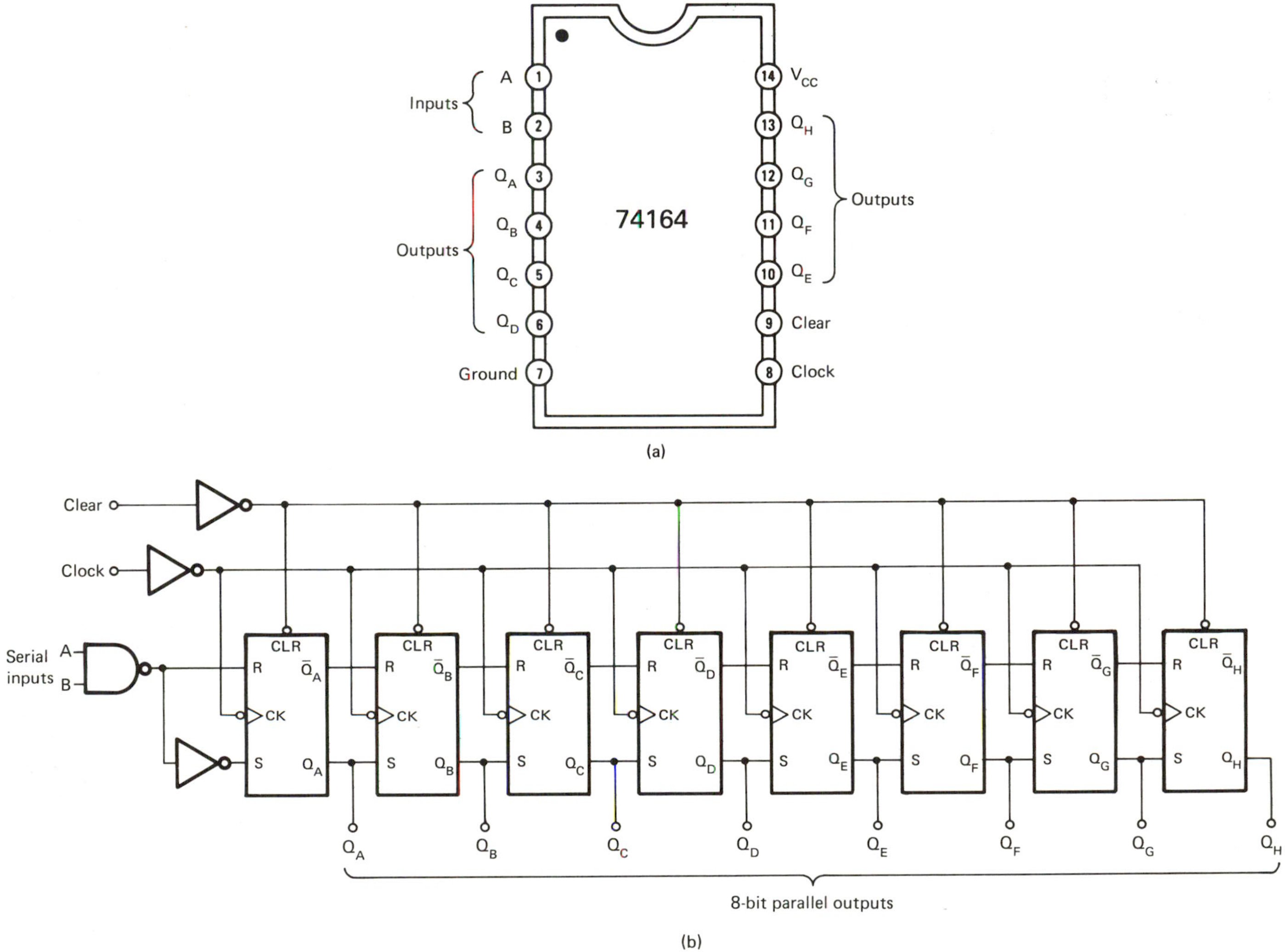

Fig. 5-8 74164 TTL SIPO shift register: (a) pin diagram, (b) logic diagram.

153

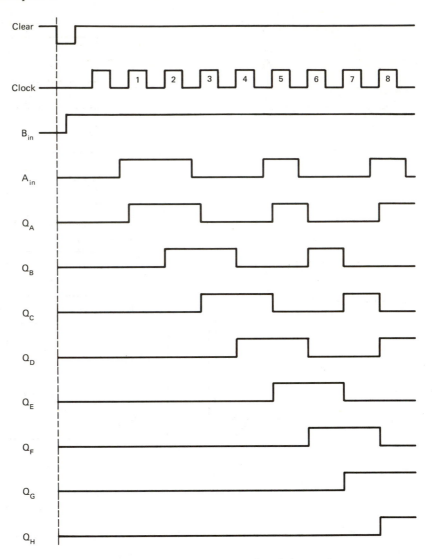

Fig. 5-9 Example of SIPO timing diagram for shifting a binary 10010011.

Virtually all integrated-circuit shift registers having parallel inputs can also be used as *serial* input types; that is, PISO and PIPO registers can be used as SISO and SIPO registers, respectively. On the other hand, registers having parallel outputs can also be used as serial output types; that is, SIPO and PIPO registers can be used as SISO and PISO registers, respectively.

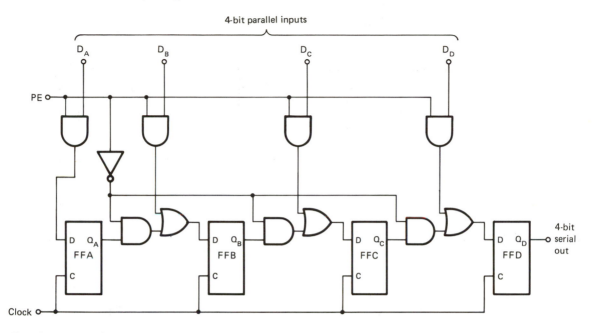

Fig. 5-10 Four-bit PISO shift register.

Right–Left and Universal Shift Registers

To this point, all the register types discussed have a *right* shift characteristic. That is, the data shift *to the right* through the flip-flop stages from input to output as the register is clocked. On the other hand, it is often useful to have the capability to shift the contents of each stage either to the right or to the left (i.e., from output to input). Computers frequently perform these shifting operations, which are

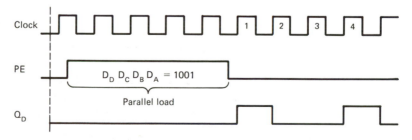

Fig. 5-11 Example of PISO shift register timing diagram for binary 1001.

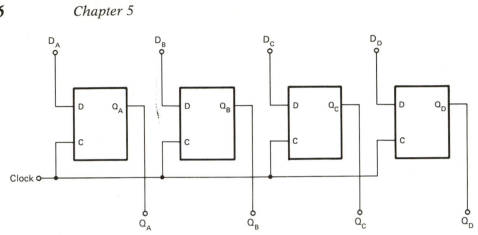

Fig. 5-12 Four-bit PIPO shift register.

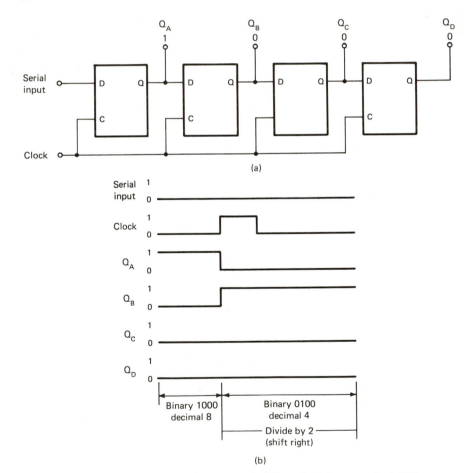

(a)

(b)

Fig. 5-13 Four-bit SIPO shift register holding the binary code 1000.

sometimes referred to as *rotate* instructions. Right shifting data one time has the same effect as *dividing* the number by 2. Left shifting a number has the same effect as *multiplying* the number by 2.

As an example, suppose we have a 4-bit SIPO register, whose stages from input to output at some given time have the binary number 1000, which is equivalent to the decimal number 8. If the serial input is 0 and a single clock pulse is applied, the output of the four stages will now be 0100, which is the decimal number 4, as shown in Figure 5–13. Therefore, when the stage contents are right shifted, the net result is a division by 2. On the other hand, if the contents of a 4-bit register are 0011 (the decimal number 3), then if the register is shifted left twice with the application of two clock pulses, the stage contents are now 1100, which is the decimal number 12. Consequently, the original contents have been twice multiplied by 2, or $3 \times 2 \times 2 = 12$.

A 4-bit shift register that has *both* a right and left shift capability, sometimes called a *bidirectional* shift register, is shown in Figure 5–14. It basically comprises four *D*-type flip-flops and four 2-input AND–OR gates. When the right/left shift (R/L) input is 1, AND gates $G1$, $G3$, $G5$, and $G7$ are enabled so that the input state of each flip-flop is *right* shifted or passed on to the next flip-flop stage each time it is clocked. If the R/L input is 0, AND gates $G2$, $G4$, $G6$, and $G8$ are enabled so that the output of each stage is directed to the input of the *preceding* stage each time it is clocked, which, in effect, *left* shifts the register contents.

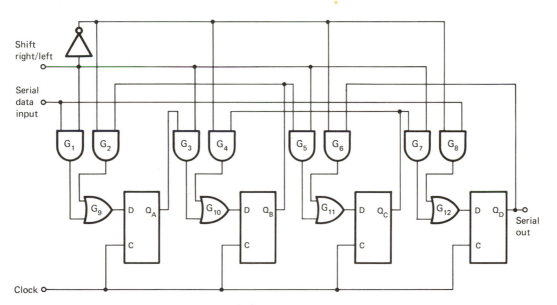

Fig. 5-14 Four-bit bidirectional SISO shift register.

If a bidirectional shift register has the capability to accept both serial and parallel data, as well as the ability to output serial data, we have what is generally referred to as a *universal shift register*. An example is the 74194 (Figure 5–15); functional CMOS equivalents are the 40104, 40194, and 74C194.

The available TTL and CMOS type shift registers are listed in Table 5–1.

BINARY RIPPLE COUNTERS

Besides creating shift registers, the flip-flop also serves as a fundamental building block in the formation of digital counters; these devices have the ability to count or total up the number of input clock pulses received. This is possible because the basic flip-flop has a memory capability, which allows it to remember the number of clock pulses that have occurred while it is waiting for the next pulse.

In Chapter 3, we discussed the most basic counter possible, the *T-type* or *toggle flip-flop*. As an example, Figure 5–16a shows a *J–K* flip-flop connected as a *T*-type. Consequently, when the clock input

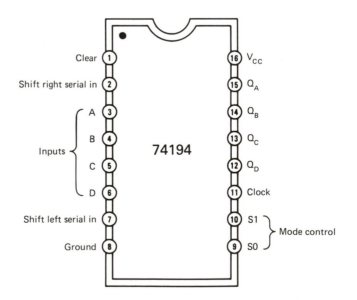

Fig. 5-15 74194 TTL universal shift register pin diagram.

Table 5-1
Available TTL and CMOS Shift Registers

	Device	Bit Length	Type	Shift Direction
TTL	7491	8	SISO	Right
	7494	4	PISO	Right
	7495	4	Universal	Both
	7496	5	Universal	Right
	7499	4	PISO	Both
	74164	8	SIPO	Right
	74165	8	PISO	Right
	74166	8	PISO	Right
	74194	4	Universal	Both
	74195	4	Universal	Right
	74198	8	Universal	Both
	74199	8	Universal	Both
CMOS	4006	Up to 18[a]	SISO	Right
	4014	6, 7, 8	PISO	Right
	4015	4[b]	SISO	Right
	4021	6, 7, 8	PISO	Right
	4031	64	SISO	Right
	4034	8	Universal	Right
	4035	4	PIPO	Right
	40100	32	SISO	Both
	40104	4	Universal	Both
	40194	4	Universal	Both
	74C95	4	Universal	Both
	74C164	8	SIPO	Right
	74C165	8	PISO	Right
	74C194	4	Universal	Both
	74C195	4	Universal	Both

[a]Bit lengths of 4, 5, 8, 9, 10, 12, 13, 14, 16, 17, and 18.
[b]Dual stages.

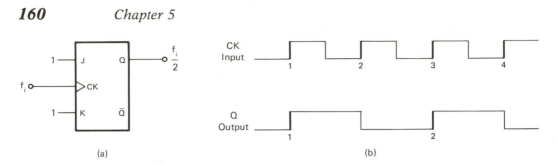

Fig. 5-16 Divide-by-2 binary counter using a *T*-type flip-flop: (a) logic diagram, (b) timing diagram.

goes from 0 to 1, output *Q* changes state. If this process is allowed to continue, we see from Figure 5–16b that we get one output cycle for every two input cycles.

Perhaps the most important feature of a *T*-type flip-flop is based on the fact that, regardless of what the duty cycle of the input wave form is, *the output will always be a 50% duty cycle pulse train,* frequently referred to as a *square wave.* In addition, *the frequency of the output square wave of a single T-type flip-flop will always be exactly one-half that of the input pulse train.* This is commonly referred to as a *divide-by-2* counter.

One useful application is to obtain a 50% duty cycle square wave from a 555 timer connected as an astable multivibrator. As discussed in Chapter 4, the duty cycle of the timer varies from a theoretical minimum of 50% to almost 100% and depends on the same two resistors used to set its free-running frequency. So, more times than not, the duty cycle of the output pulse train will not be 50%. A simple solution is to connect a *T*-type flip-flop to the output of the timer so that the output of the flip-flop will have a 50% duty cycle, although the output frequency will be *one-half* that of the timer. To overcome this, *double* the output frequency of the timer to give the proper output frequency from the flip-flop.

With the *T*-type flip-flop shown in Figure 5–16, we have only the ability to count either 0 or 1. That is, when output *Q* goes from 0 to 1 (i.e., the flip-flop is set) after a single 0-to-1 or positive transition of the clock pulse, the flip-flop then indicates that it has counted a single input pulse. After a second positive transition, the *Q* is then 0 (i.e., reset). If the flip-flop is clocked on the 1-to-0 or negative transition of the clock pulse, output *Q* then changes state each time a negative transition has occurred, as shown in Figure 5–17. Note that an *inversion circle* is used on the logic symbol for the clock input to indicate that the flip-flop is toggled on the negative transition. For both arrangements, however, the binary number for the decimal number 2 is 10, not 0. Consequently, this single flip-flop stage is only capable of knowing if no clock pulses have occurred ($Q = 0$) or if only a single clock pulse has occurred, in which case, $Q = 1$. If additional input pulses occur, the output of the flip-flop either indicates only a 0, if an even number of clock pulses have occurred, or a 1, indicating that an odd number of clock pulses have been counted. However, in both cases we do not know how many.

To increase the counting capacity, we then connect a string of *T*-type flip-flops in cascade, such as the 3-stage counter shown in Figure 5–18a. The *Q* output of a flip-flop stage is connected to the toggle input of the next, forming what is generally referred to as a *divider string*. Since each flip-flop functions as divide-by-2 circuit, the output frequency of each flip-flop is one-half its input. Consequently, the output of *FFA* is $f_i/2$; the output of *FFB* is one-half the output of *FFA*, or $f_i/4$; and the output of *FFC* is one-half the output of *FFB*, or $f_i/8$. In effect, the output frequency decreases by one-half each time an additional toggle flip-flop stage is added. In general, the output frequency after *N* *T*-type flip-flops is

$$f_o = \frac{f_i}{2^N} \qquad (5–1)$$

The integer number equal to the quantity 2^N is known as the *modulus* of the counter, which is thus sometimes referred to as a *mod-N* or *modulo-N* counter. Therefore, the three-stage counter of Figure 5–18a is a mod-8, or modulo-8, or divide-by-8 counter as the output frequency has one output cycle for every eight input cycles.

If we carefully inspect the timing diagram of Figure 5–18b, we notice that there are three output signals in addition to the input clock signal. These are the outputs of the individual flip-flop stages. Assuming initially that all three flip-flops are reset, the *Q* outputs are all 0, or, in

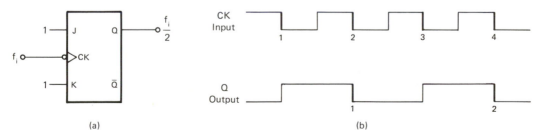

(a)   (b)

Fig. 5–17 Divide-by-2 binary counter clocked on the negative trailing edge: (a) logic diagram, (b) timing diagram.

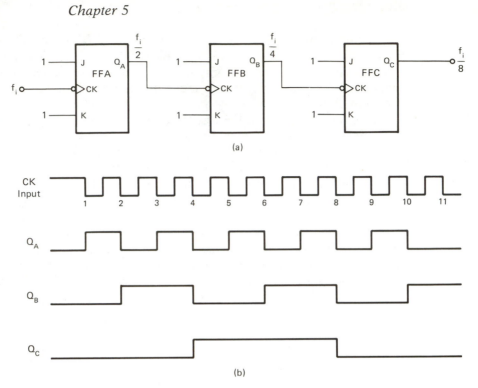

(a)

(b)

Fig. 5-18 Three-stage binary ripple up counter: (a) logic diagram, (b) timing diagram.

order, *CBA* = 000. This is equivalent to the decimal number 0. After the first *negative* clock pulse, *FFA* is set, so outputs *CBA* = 001, or the decimal number 1. Therefore, we now know that a single clock pulse has occurred and the three stage counter has "counted" it. After the second negative clock pulse has occurred, *FFA* is reset while *FFB* is set, so that outputs *CBA* = 010, or the decimal number 2. On the third clock pulse, *FFA* is again set while *FFB* remains set, so that outputs *CBA* = 011, which tells us that this counter has now counted three clock pulses, since the 3-bit binary number 011 is equal to the decimal number 3. Consequently, the output states of the flip-flop stages are counting *upward in binary code* from 000 to 111, which is the decimal number 7. Since 0 is a valid number in the count sequence from 0 to 7, eight combinations of output states are possible, hence the name *modulus-8* counter. The eight output states are summarized in Table 5–2.

Since the output pulse train of the last flip-flop stage has a frequency of one-eighth that of the input, it is also correct to call this circuit a divide-by-8 counter. Regardless of the specific name we attach

Table 5-2
Output States for a
Modulus-8 Up Counter

Input Clock Pulse	Outputs		
	FFC	*FFB*	*FFA*
0	0	0	0
1	0	0	1
2	0	1	0
3	0	1	1
4	1	0	0
5	1	0	1
6	1	1	0
7	1	1	1

to this circuit, it produces an *upward* binary count from 0 to 7. Because the counting process is performed in binary, such an arrangement of cascaded toggle flip-flops is generally referred to as a *binary* or *ripple* counter. The term "ripple" refers to the fact that the 1s and 0s "ripple" down the divider chain from the input stage to the output, just as in a shift register. Since the flip-flop stages are not made to change states at exactly the same time, as the clock pulses are not directly connected to the toggle or clock input of each flip-flop, a binary ripple counter is frequently referred to as an *asynchronous binary up counter*.

Such a modulo-8 binary ripple counter is used in the 7493, 74293, and 74C93 4-bit counters, as shown in Figure 5–19. Although there are four stages, three flip-flops are grouped together with their own input and output connections as divide-by-8 or modulo-8 binary ripple connections. In addition, a single flip-flop stage is used as a divide-by-2 counter. The reset inputs of each stage are tied together and activated by a 2-input NAND gate. If reset inputs $R_{0(1)}$ and $R_{0(2)}$ are both at 1, then all four stages are simultaneously reset. If either or both reset inputs are 0, then normal binary counting occurs.

By connecting the output of the divide-by-2 counter to the input of the divide-by-8 counter, we have a *divide-by-16* binary ripple counter. When there are two separate counter sections, such as the 7493/74C93/74293 devices, they are said to be organized in a *2-by-8 or 2 × 8* format, since they contain both divide-by-2 and divide-by-8 counters.

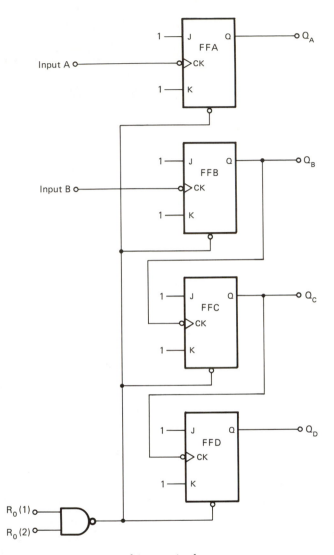

Fig. 5-19 Four-stage binary ripple up counter.

In general, when a modulus-X counter is cascaded with a modulus-Y counter, the resultant effect is a counter whose modulus is the *product of X and Y.* In the case of the 7493 and the like, a modulus-2 counter cascaded with a modulus-8 counter results in a counter having a modulus of 2 times 8, or 16. In theory, it does not make any difference whether the modulus-8 section is placed ahead of the modulus-2 section in the divider chain or if the reverse is true. Both result in a modulus-16 counter. In actual practice, however, the counter section with the smaller modulus is placed *first* (i.e., the modulus-2 section), as its maximum input frequency rating is higher than the larger section.

To illustrate this point, the maximum input frequency that the modulus-2 or *A* counter of the 7493 can handle is typically 32 MHz, while the maximum input frequency of its three-stage modulus-8 or *B* counter is typically 16 MHz. If the input frequency is always less than 16 MHz, then it makes no difference which section is placed first in the divider chain as either counter stage is capable of toggling at rates up to 16 MHz. On the other hand, if the input frequency is greater than 16 MHz but less than 32 MHz, the "faster" modulus-2 stage must be placed ahead of the "slower" 16 MHz, modulus-8 stage. Consequently, the resultant divider chain is then only as strong as its weakest (i.e., 16 MHz) link.

Instead of using *N* number of flip-flops to create a modulus-2^N binary ripple counter, a number of TTL and CMOS devices are available (see Table 5–3) to fill most needs. For higher modulos, such as 256, we merely cascade two mod-16 counters, as illustrated in Figure 5–20.

So far we have shown how the binary ripple counter counts upward. To count *downward*, we simply clock the next stage from the $\overline{Q}$ output of the previous stage, as shown in Figure 5–21; the output states are summarized in Table 5–4.

SYNCHRONOUS COUNTERS

In contrast to the binary ripple or asynchronous counter, where each flip-flop stage is clocked or toggled by the previous flip-flop stage, a *synchronous binary* counter has all its flip-flop stages clocked *at the same time*. As shown by the two-stage synchronous counter of Figure 5–22a, the CLOCK inputs of both *J–K* flip-flops are tied together.

Initially, let's assume that both flip-flops are reset so that $Q_A = Q_B = 0$, which is equal to the decimal number 0. *FFA*, connected as a *T*-type, first toggles when the 0-to-1 transition of the first clock pulse (CK_1) is applied. At this point, $Q_A = 1$ and $Q_B = 0$, which is 01 in binary and equals the decimal number 1. Although the *J* and *K* inputs of *FFB* are now both 1, which would normally cause it to toggle when a clock pulse is applied, output Q_A did not *immediately* change from 0 to 1 during the first clock pulse. Instead, Q_A changed from 0 to 1 a finite amount of time later, caused by the propagation delay through *FFA*. At the 0-to-1 transition

Table 5-3
TTL and CMOS Binary Ripple Counters

		Stages	Modulus	Organization	Clocking	Typical Max. Frequency (MHz)	Remarks
TTL	7493	4	16	2×8	Negative	32, 16	
	74293	4	16	2×8	Negative	32, 16	
CMOS	74C93	4	16	2×8	Negative	2, 5[a]	
	4020	14	16, 384	—	Negative	7, 16, 24[b]	
	4024	7	128	—	Negative	7, 16, 24[b]	No outputs for mod-4, 8
	4040	12	4096	—	Negative	7, 16, 24[b]	
	4045	21	2,097,152	—	Negative	7, 16, 24[b]	
	4060	14	16, 384	—	Negative	7, 16, 24[b]	No outputs for mod-2, 8, 2048

[a]For supply voltages of 5 and 10 V.
[b]For supply voltages of 5, 10, and 15 V.

166

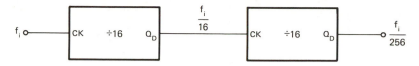

Fig. 5-20 Cascading two divide-by-16 counters to create a divide-by-256 counter.

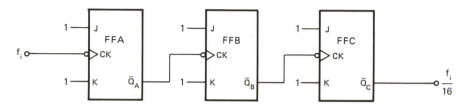

Fig. 5-21 Three-stage binary ripple down counter.

Table 5-4
Output States for a Modulus-16 Down Counter

Input Clock Pulse	Outputs			
	FFD	*FFC*	*FFB*	*FFA*
0	0	0	0	0
1	1	1	1	1
2	1	1	1	0
3	1	1	0	1
4	1	1	0	0
5	1	0	1	1
6	1	0	1	0
7	1	0	0	1
8	1	0	0	0
9	0	1	1	1
10	0	1	1	0
11	0	1	0	1
12	0	1	0	0
13	0	0	1	1
14	0	0	1	0
15	0	0	0	1
16	0	0	0	0

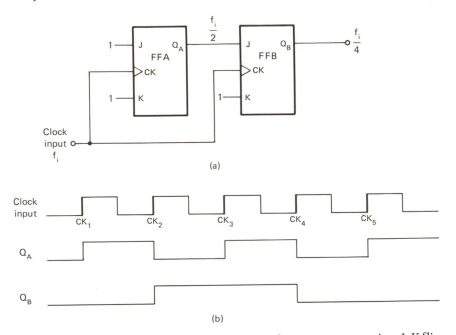

Fig. 5-22 Two-stage (modulo-4) synchronous binary counter using *J–K* flip-flops: (a) logic diagram, (b) timing diagram.

of the second clock pulse (*CK₂*), *FFA* again toggles and $Q_A = 0$. Since J_B and K_B are both 1 at the start of *CK₂*, *FFB* then toggles before the propagation-delayed pulse can have any effect on *FFB*. Consequently, $Q_B = 1$, which is 10 in binary and equals the decimal number 2. This makes sense, as we have just now completed two input clock pulses.

At the 0-to-1 transition of the third clock pulse, *FFA* again toggles, causing it to be set, so $Q_A = 1$. Since the *J* and *K* inputs of *FFB* are briefly 0 before going to 1 when *CK₃* is applied due to propagation delay, *FFB* remains set ($Q_B = 1$) so that both flip-flops indicate the binary number 11, which is the decimal number 3. When the fourth clock pulse is applied to *FFA*, it is then reset, and since the *J* and *K* inputs of *FFB* are briefly 1 during the fourth clock pulse, it too is reset. Consequently, $Q_A = Q_B = 0$, which equals the decimal 0 and is the original state we started with. The sequence of events for all four clock pulses is illustrated by the timing diagram of Figure 5–22b. It should be noted that, although there are finite propagation delays associated with *FFA*, the timing sequences are drawn without them, as they are generally omitted to keep things simple.

Figure 5–23 shows a three-stage synchronous binary up counter and its associated timing diagram. As before, *FFA* is connected as a simple *T*-type and is repeatedly toggled each time the clock pulse goes from 0 to 1. If initially all flip-flops are reset, then following the first clock pulse $Q_A = 1$, $Q_B = 0$, and $Q_C = 0$, which equals the decimal number 1. When the second clock pulse occurs, *FFA* resets, while *FFB* is set. Because of the propagation delay of *FFA*, the *J* input is momentarily 0 while $Q_A = 1$. These two signals feed the AND gate, whose output is then 0 at the moment CK_2 occurs. Since the *J* and *K* inputs are both 0 at this point, *FFC* remains reset and $Q_C = 0$.

Just prior to CK_3, $Q_A = 0$, $Q_B = 1$, and $Q_C = 0$ (binary 010 equals the decimal number 2). When *FFA* receives the third clock pulse, it again is set ($Q_A = 1$). Before the pulse can finish propagating through *FFB*, Q_B is still 1, so the output of the AND gate and the *J–K* inputs of *FFC* are 0, causing *FFC* to remain reset ($Q_C = 0$). After CK_3 propagates through *FFB*, $Q_A = 1$, $Q_B = 1$, and $Q_C = 0$, indicating the decimal number 3. From the truth table of Table 5–5 and the timing diagram of Figure 5–23b, we see that five more clock pulses are required before all three flip-flops

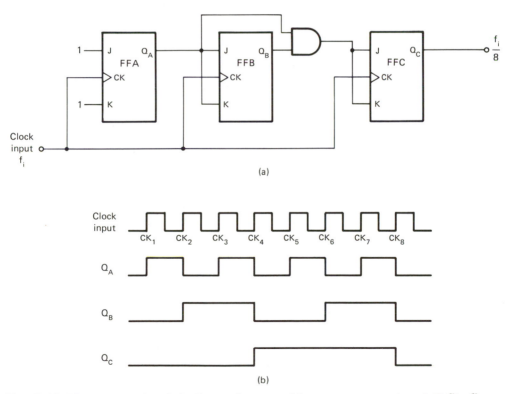

Fig. 5-23 Three-stage (modulo-8) synchronous binary counter using *J–K* flip-flops: (a) logic diagram, (b) timing diagram.

Table 5-5
Output States for a Three-Stage
Synchronous Binary Up Counter

Input Clock Pulse	Q_C	Outputs Q_B	Q_A
0	0	0	0
1	0	0	1
2	0	1	0
3	0	1	1
4	1	0	0
5	1	0	1
6	1	1	0
7	1	1	1
8	0	0	0
9	0	0	1
10	0	1	0
11	0	1	1
12	1	0	0
13	1	0	1
	etc.		

are reset, which is our initial condition. Figure 5–24 shows how to form a four-stage synchronous binary counter. If all four flip-flops are initially reset, then 16 clock pulses are required to again reset all four flip-flop stages.

Like the binary ripple counter, it is also possible to construct synchronous binary *down* counters by simply using the complementary output. Therefore, the three-stage up counter of Figure 5–23 now becomes the down counter shown in Figure 5–25 by feeding the next flip-flop stage with the $\overline{Q}$ output of the previous stage.

Both ripple and synchronous binary counters have the same end result. That is, a single output pulse is generated for every 2^N input clock pulses, where N is the number of flip-flop stages that make up the counter. If both types give the same result, does one have any advantage over the other? Since one stage clocks the next, the

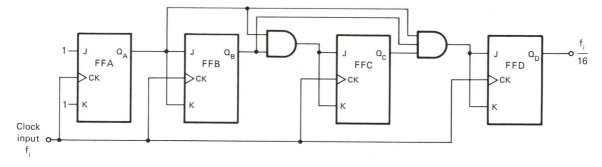

Fig. 5-24 Four-stage (modulo-16) synchronous binary counter using *J–K* flip-flops.

propagation delays associated with a ripple counter tend to build up as the number of stages is increased. Consequently, the output states of all the stages will not be true until the counter has completely settled down, that is, until the clocking pulse has passed through every stage of the counter. Until this occurs, invalid intermediate output states exist, which may erroneously drive other circuits. Since synchronous binary counter stages are clocked simultaneously, we eliminate the pitfalls associated with ripple counters. However, we pay a price. In return for their faster speed, synchronous counters generally require more logic elements, resulting in increased power consumption and cost.

Using binary counters, whether synchronous or asynchronous, it is possible to create a counter having virtually any modulus. The following sections will focus on those TTL and CMOS counters intended for specific applications.

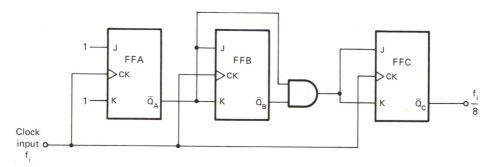

Fig. 5-25 Three-stage synchronous binary down counter.

DECADE COUNTERS

A *decade* counter has ten states in its counting sequence. That is, it produces a single output pulse for every ten input pulses. As we shall see later, decade counters are tremendously important devices that are required for the building of digital time clocks, frequency counters, and other functions requiring digital displays.

Figure 5–26 shows a typical block and timing diagram for a decade counter. There is a single clock input and four *binary coded decimal* (BCD) outputs. Table 5–6 shows the BCD decade counting sequence. The BCD decade counting sequence follows exactly the binary counting sequence for the decimal numbers 0 through 9. After

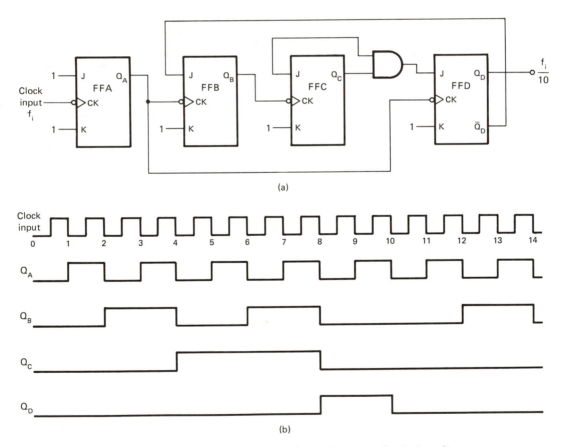

Fig. 5-26 BCD decade (modulo-10) counter: (a) logic diagram, (b) timing diagram.

Table 5-6
Output States for a
BCD Decade Counter

Clock Cycle	Outputs			
	Q_D	Q_C	Q_B	Q_A
0	0	0	0	0
1	0	0	0	1
2	0	0	1	0
3	0	0	1	1
4	0	1	0	0
5	0	1	0	1
6	0	1	1	0
7	0	1	1	1
8	1	0	0	0
9	1	0	0	1
10	0	0	0	0
11	0	0	0	1
12	0	0	1	0
13	0	0	1	1
14	0	1	0	0
		etc.		

the tenth clock cycle, the counter recycles to the 0 state. It should be noticed that, although the binary output count of the decimal number 10 is not specifically mentioned, the decade counter actually has ten separate and unique output states. Starting with 0 as the first possible state, we then count all the way up to 9, which is the tenth and last possible state. Further inspection of Table 5–6 shows that outputs Q_C and Q_D will repeat themselves every ten input cycles. For example, output Q_C goes from 0 to 1 on the 4th and 14th clock cycles. If we were to extend this table further, we would find that output Q_D goes from 0 to 1 at the 8th and 18th, 28th, and so on, clock cycles.

The most common TTL decade counter is the 7490 (and the equivalent CMOS 74C90), whose pin and logic diagrams are shown in Figure 5–27. The 7490 is made up of four flip-flops and additional gates

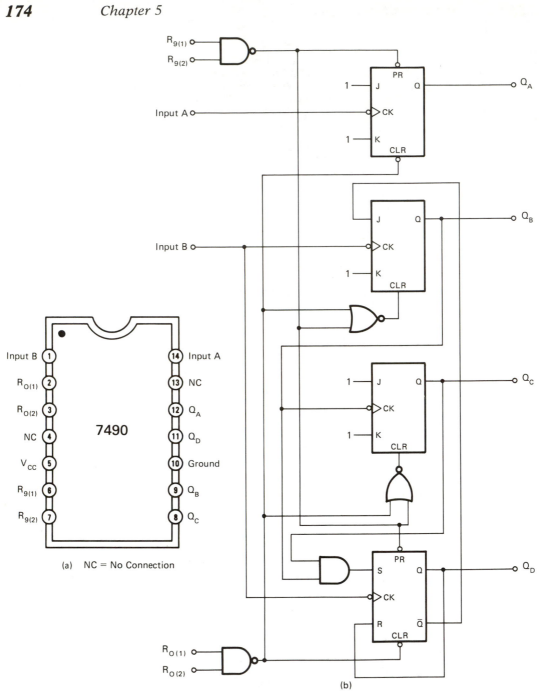

Fig. 5-27 7490 TTL decade counter: (a) pin diagram; (b) logic diagram.

to provide a divide-by-2 binary counter and a three-stage binary counter where the count cycle gives a divide-by-5. In addition to the four BCD outputs, there are two gated RESET-to-0 inputs, $R_{0(1)}$ and $R_{0(2)}$, and two gated RESET-to-9 inputs, $R_{9(1)}$ and $R_{9(2)}$. When logic 1s are applied to both RESET-to-0 inputs (pins 2 and 3), the 7490 counter is immediately reset to 0 so that all four outputs are 0. On the other hand, when logic 1s are applied to both the RESET-to-9 inputs (pins 6 and 7), the counter is immediately set to the 9 count so that $Q_D=1$, $Q_C=0$, $Q_B=0$, and $Q_A = 1$.

To wire the 7490 as a BCD counter, the Q_A output is tied to the B input (pin 1) while the clock input is applied to the A input (pin 14), as shown in Figure 5–28. This arrangement gives the outputs that follow the timing diagram of Figure 5–26b and the sequence of Table 5–6. By connecting the four BCD outputs to a LED display (see Chapter 7), we are then able to "see" how many clock pulses have been counted.

Unfortunately, with a single 7490 decade counter, we have the capability to count numbers only from 0 to 9. To count higher numbers, such as those from 10 to 99, we need to cascade an additional 7490 BCD counter by connecting the Q_D output of the first stage to the A input of the next counter stage, as shown in Figure 5–29. In this manner, assuming that both counters are initially reset to 0, decade counter $U1$ (i.e., the *units* decade) counts from 0 to 9 as before. On the tenth clock cycle, counter $U1$ resets its count to 0 and counter $U2$ (i.e., the *tens* decade) counts a 1, which is the decimal number 10. Counter $U2$ will then increase its count by 1 every time counter $U1$ goes past the count of 9. In this manner, we can cascade as many decade counters as we

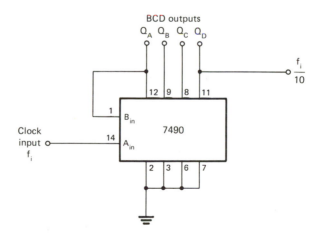

Fig. 5-28 7490 decade counter with BCD output.

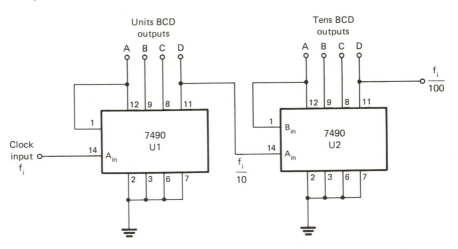

Fig. 5-29 Two-decade counter with BCD outputs using a pair of cascaded 7490 counters.

wish. If six BCD decade counters were cascaded in a manner similar to the two-counter circuit of Figure 5–29, we would then be able to count up to 10^6, or 1 million clock cycles.

Instead of using the 7490 as a BCD counter, where we are able to display the count sequence, we can use the 7490 simply as a divide-by-10 counter, producing a single output cycle for every ten input clock cycles. As explained earlier, when connected as a BCD counter, outputs Q_C and Q_D repeat their sequences every ten clock cycles. However, neither of these outputs is *symmetrical*, so the output would be 0 for five consecutive cycles and 1 for five consecutive cycles. From the timing diagram of Figure 5–26b, we see that output Q_C is 1 for four consecutive cycles and 0 for six cycles, while output Q_D is 1 for only 2 consecutive cycles and 0 for eight. In some circumstances, however, the output of a counter must be symmetrical. Using the 7490, we can do this by tying the Q_D output to the A input and applying the clock signal to the B input. The resulting divide-by-10 output is taken from Q_A (the divide-by-2 counter section), as shown in Figure 5–30.

Figure 5–31 illustrates how three 7490 counters are cascaded to divide the input clock frequency by 1000 so that the output cycle is symmetrical. As with the BCD configuration, any number of 7490 decade counters may be cascaded in this manner.

Besides the CMOS equivalent 74C90, the 4518 (Figure 5–32) is a dual synchronous decade counter consisting of a pair of identical four-stage counters, each with interchangeable CLOCK and ENABLE inputs. This feature allows the counter section to be clocked on positive

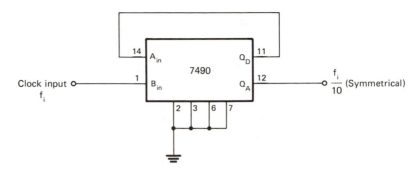

Fig. 5-30 7490 modulo-10 counter with symmetrical output.

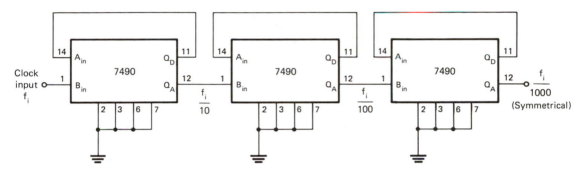

Fig. 5-31 Modulo-1000 counter.

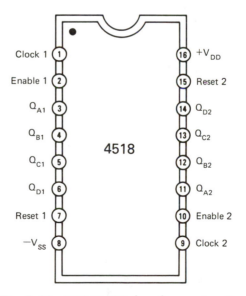

Fig. 5-32 4518 CMOS decade counter pin diagram.

or negative transitions, as shown in Figure 5–33 and summarized in Table 5–7.

The 4518 can be cascaded in the *ripple* mode connecting the Q_D output to the ENABLE input of the next 4518 counter, giving a divide-by-100 counter as shown in Figure 5–34. As in the ripple counters discussed earlier, propagation delay increases for each counter added. Figure 5–35 shows how two 4518s can be *synchronously* cascaded with the aid of an AND gate.

Although not a decade counter, the 7492 divide-by-12 TTL counter is worth mentioning here because it is frequently used in conjunction with the 7490. As shown in the pin and logic diagrams of Figure 5–36, the 7492, like the 7490, is made up of four flip-flops and additional gates to provide both a divide-by-2 binary counter and a three-stage binary counter where the count cycle gives a divide-by-6. In addition to the four BCD outputs, there are two gated RESET-to-0 inputs, $R_{0(1)}$ and $R_{0(2)}$.

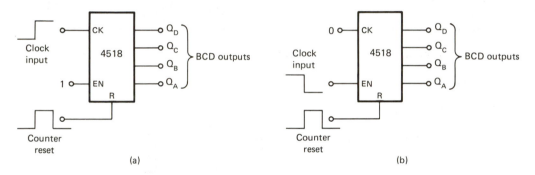

Fig. 5-33 4518 decade counter with BCD output: (a) positive edge clocking, (b) negative edge clocking.

Table 5-7
4518 Truth Table

CLOCK Input	ENABLE Input	Result
⌐	1	Count up
0	⌐	Count up
⌐	0 or 1	No change
0 or 1	⌐	No change
⌐	0	No change
1	⌐	No change

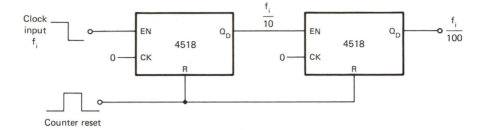

Fig. 5-34 Ripple cascading of two 4518 counters.

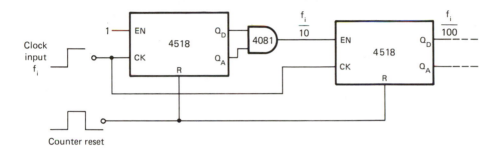

Fig. 5-35 Synchronous cascading of two 4518 counters.

Figure 5–37 shows the 7492 wired to divide the input frequency by 6. When cascaded with a 7490 divide-by-10 counter, we can divide an input frequency by 60. Many simple digital time clocks use this principle to provide a 1-Hz clock signal from the 60-Hz power line frequency, as shown in Figure 5–38. This is as accurate as any conventional motor-driven clock, since they both get their signals from the 60-Hz power line. Figure 5–39 shows the 7492 wired to divide by 6, with 3-bit binary outputs. In this arrangement, we can count from 0 to 5, repeating the cycle as many times as needed, as summarized by Table 5–8.

A simple digital time clock counting minutes and seconds can be built using 7490 and 7492 counters as shown in Figure 5–40. As before, counters $U1$ and $U2$ divide the 60-Hz power line frequency by 60 down to 1 pulse each second. Counters $U3$ and $U4$ count the output pulses from $U2$, which is the elapsed time in seconds from 00 to 59, and repeat this every 60 clock cycles, or every 60 seconds. Counters $U5$ and $U6$ function the same as counters $U3$ and $U4$, but this time they count the time in minutes. Since there are 60 seconds in every minute, when counters $U4$ and $U3$ both reset to zero after every 60 input clock pulses,

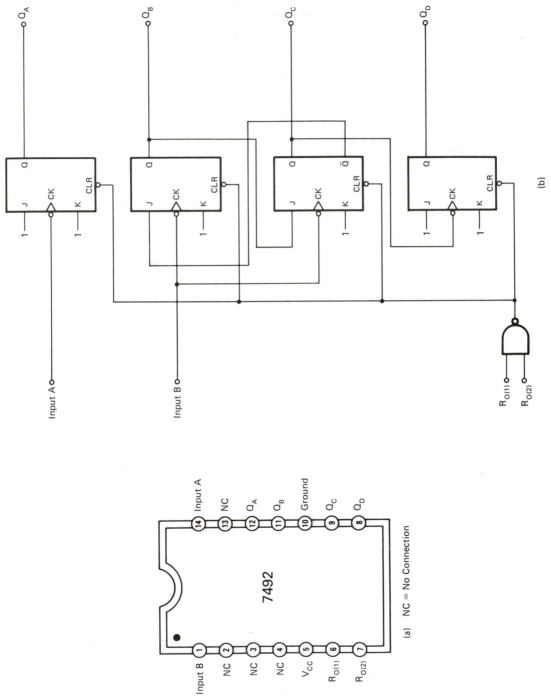

Fig. 5-36 7492 TTL divide-by-12 counter: (a) pin diagram, (b) logic diagram.

(a) NC = No Connection

(b)

180

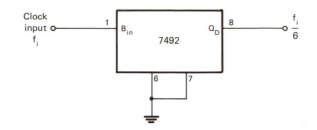

Fig. 5-37 7492 divide-by-6 counter.

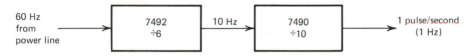

Fig. 5-38 Dividing the 60-Hz power line frequency to 1 Hz.

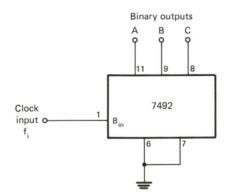

Fig. 5-39 7492 divide-by-6 counter with 3-bit binary output.

Table 5-8
Output States for a 7492 Divide-by-6 Counter

Clock	Outputs		
Cycle	Q_C	Q_B	Q_A
0	0	0	0
1	0	0	1
2	0	1	0
3	0	1	1
4	1	0	0
5	1	0	1
6	0	0	0
7	0	0	1
8	0	1	0
		etc.	

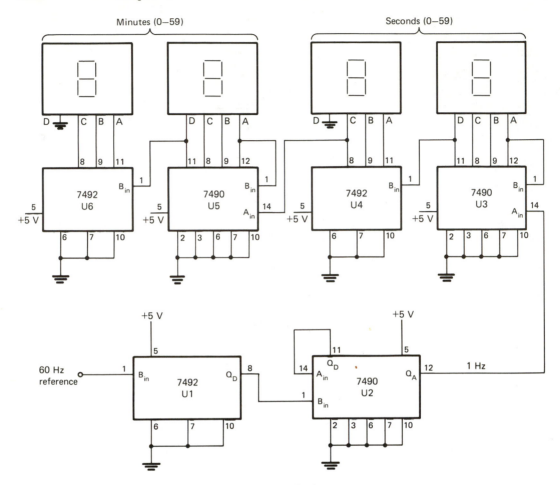

Fig. 5-40 A 59-minute, 59-second elapsed time clock.

counter $U5$ increments its count by 1, and so on, until the minutes count reaches 59.

If counters $U3$ through $U6$ were initially reset to display 00:00, then after 60×60, or 3600 clock cycles (3600 seconds), each counter would be automatically reset to zero, and the counting process starts over again. In this example, the circuitry required for displaying the BCD output of the counter stages has been omitted. Specific decoder circuitry and digital displays are discussed in Chapter 7, as well as a complete digital time clock circuit.

PROGRAMMABLE COUNTERS

It is possible to decode the outputs of various TTL and CMOS counters to give a modulus different from that normally expected. For example, a 7490 decade counter, in addition to being able to divide clock inputs by 2, 5, and 10, can be arranged to divide also by 6, 7, and 9. However, these arrangements generally have their shortcomings, such as lacking the ability to have the binary count externally reset back to zero. Instead of discussing how to create a variable modulus counter with devices normally intended for fixed applications, this section discusses those counters that have the specific ability to be *programmed*. Counters of this type, generally referred to as *programmable counters*, can be told to count either up or down, as well as have their modulus programmed or *preset* to a given value using the standard 4-bit BCD or binary codes.

An up/down or *bidirectional* counter is one whose counting sequence can either increase or decrease. In addition, its count direction may be changed at any time. As an example, suppose an up/down counter starts counting up from 0 to 6 and then it is told to count down three counts, and then again count upward. Such a sequence would appear as

$$\underbrace{0,\ 1,\ 2,\ 3,\ 4,\ 5,\ 6,}_{\text{up}}\ \underbrace{5,\ 4,\ 3,}_{\text{down}}\ \underbrace{4,\ 5,\ 6,\ 7,\ .\ .\ ,\ \text{etc.}}_{\text{up}}$$

Since up/down counters are able to add to as well as subtract from the stored count, they are often called *add/subtract* counters. Rather than analyzing the logic of a typical up/down counter, we shall look at several TTL and CMOS devices designed for this purpose.

74192 TTL Up/Down Decade Counter

The 74192, whose pin diagram is shown in Figure 5–41, is a 4-bit BCD type whose modulus can be programmed, or *preset*, to a given value using the standard BCD code at its DATA inputs. Alternative names for these inputs are *preset*, *jam*, and *load*.

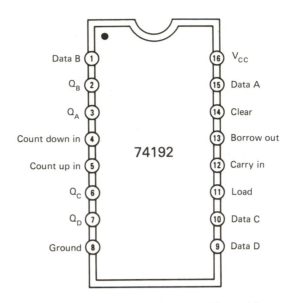

Fig. 5-41 74192 TTL programmable up/down counter pin diagram.

When the LOAD input is briefly 0, the BCD code present at the DATA inputs are *parallel loaded* to determine the counter's modulus. For example, if the BCD code present at the four DATA inputs is *DCBA* = 1000, then the modulus of the counter is preset at 8 when the LOAD input is momentarily 0. The 74192 advances one count on each positive leading edge at the COUNT UP input and decreases by one count when the clock is applied to the COUNT DOWN input. When using one of the two count inputs, the remaining input is held at logic 1.

Figure 5–42 illustrates the general scheme for cascading three 74192 counters so that the overall circuit has a modulus from 1 to 999. For example, if the DATA inputs for the *units* decade (*U1*) are set at $DCBA_1 = 0011$, the *tens* decade (*U2*) set at $DCBA_{10} = 1001$, and the *hundreds* decade (*U3*) at $DCBA_{100} = 0111$, then the resultant counter would have a maximum count of

$$(7 \times 10^2) + (9 \times 10^1) + (3 \times 10^0) = 793$$

4510 CMOS Up/Down Decade Counter

The 4510 up/down decade counter, whose pin diagram is shown in Figure 5–43, is similar in operation to the 74192 TTL device. Unlike the

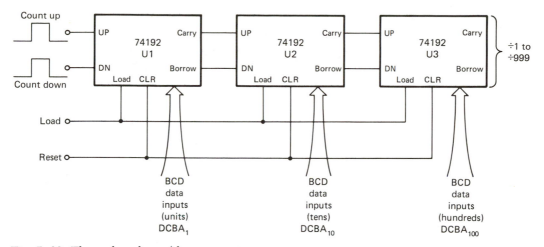

Fig. 5-42 Three-decade up/down counter.

74192 (or 74C192), the 4510 does not have separate inputs for counting up or down, but rather a single UP/DOWN input to determine if the count is up or down on each positive leading edge of the clock. If the UP/DOWN input is 1, the count is upward; a 0 causes the count to move downward. The truth table for the 4510 is summarized in Table 5–9.

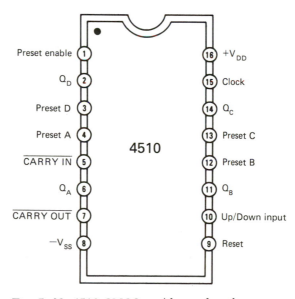

Fig. 5-43 4510 CMOS up/down decade counter pin diagram.

Table 5-9
Truth Table for the 4510 Up/Down Decade Counter

CLOCK Input	CARRY IN Input	UP/DOWN Input	PRESET ENABLE	RESET Input	Result
0 or 1	1	0 or 1	0	0	No count
⌐	0	1	0	0	Count up
⌐	0	0	0	0	Count down
0 or 1	0 or 1	0 or 1	1	0	Preset with BCD code
0 or 1	0 or 1	0 or 1	0 or 1	1	Reset to 0

As shown in Figure 5–44a, two or more 4510 counters can be cascaded for *parallel* clocking by connecting the $\overline{\text{CARRY OUT}}$ (*CO*) signal of the first stage (*U1*) to the $\overline{\text{CARRY IN}}$ (*CI*) pin of the next stage as the input signal simultaneously drives the CLOCK input of all stages. For *ripple* clocking, the CLOCK and *CI* inputs of each stage are connected together (Figure 5–44b). The BCD code is *parallel loaded* into the counter by momentarily setting the PRESET ENABLE input at logic 0.

SHIFT REGISTER COUNTERS

At the beginning of this chapter we discussed the shift register. In addition to shifting a 0 or 1 through successive flip-flop registers each time it is clocked, the basic shift register can be arranged to form several types of clocked counters. As opposed to the basic shift register, all *shift register counters* use a feedback connection whereby the output of the last flip-flop stage is connected back to the first flip-flop stage. Shift register counters are divided into two broad categories, *ring counters* and *Johnson counters*, which are discussed separately in the following sections.

Ring Counter

The basic ring counter, sometimes referred to as a *circulating shift register*, shown in Figure 5–45a using *D*-type flip-flops, has the *Q* out-

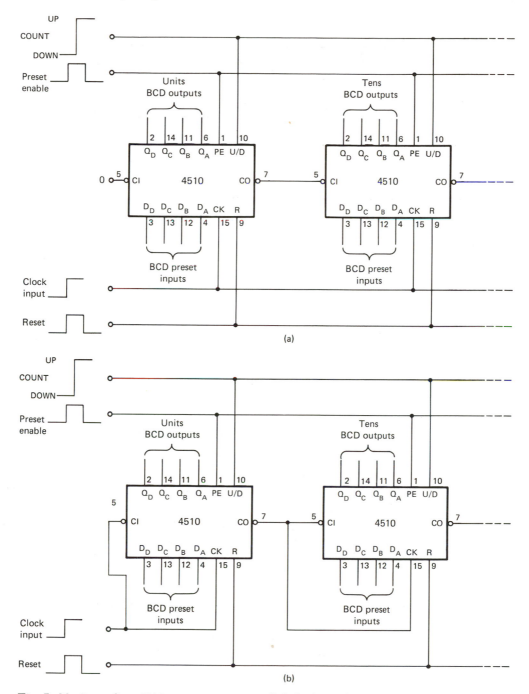

Fig. 5-44 Cascading 4510 counters: (a) parallel clocking, (b) ripple clocking.

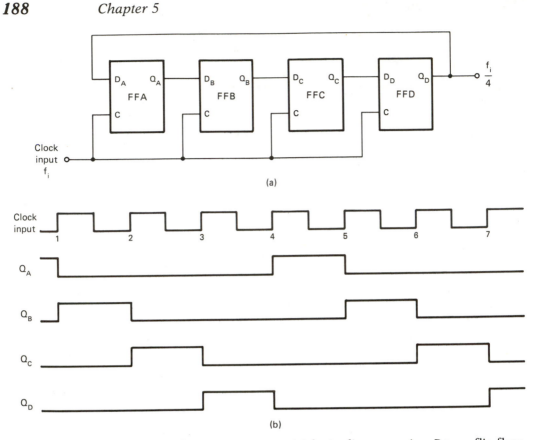

(a)

(b)

Fig. 5-45 Four-stage (modulo-4) ring counter: (a) logic diagram using *D*-type flip-flops, (b) timing diagram.

put of the last flip-flop stage connected to the *D* input of the first flip-flop. In this manner, the information shifts from left to right and back around, starting the process all over again. Usually a logic 1 is placed only in one stage and made to "circulate" around the register for as long as clock pulses are present, as shown by the associated timing diagram in Figure 5–45b.

It should be noted that, since there are four stages in this example, the output frequency is one-fourth the input clocking frequency, or a divide-by-4 counter. Also, the output has a duty cycle that is 1/4, or 25%. In general, we can state that, for a ring counter having *N* flip-flop stages, the frequency and percent duty cycle of the resultant output are

$$f_o = \frac{f_i}{N} \qquad (5-2)$$

$$\% \text{ Duty cycle} = \frac{1}{N} \times 100 \qquad\qquad (5\text{--}3)$$

We are not restricted to using *D*-type flip-flops in building ring counters. Figure 5–46 shows a four-stage ring counter using *J–K* flip-flops. However, since the *J* and *K* inputs of each stage are complementary, each *J-K* flip-flop is, in effect, connected as a *D*-type flip-flop.

From Eq. 5–2, ring counters for any modulus can be constructed; the modulus simply equals the number of flip-flop stages. As a general rule, however, a ring counter requires more flip-flop stages than a binary counter having the same modulus. A divide-by-16 *binary* counter requires only four *T*-type flip-flops, while we would need 16 *D*-type flip-flops to build a divide-by-16 *ring* counter.

As mentioned at the beginning of this section, ring counters must be initialized so that *only one of the flip-flops is set (Q = 1)*. When power is first applied to the ring counter circuit of either Figure 5–45 or 5–46, there is only a small chance that only one flip-flop will achieve this condition. Consequently, the flip-flops used should have both **PRESET** (SET) and **CLEAR** (RESET) inputs. Using both *D*-type and *J-K* flip-flops, Figure 5–47 shows how a four-stage ring counter is initialized with the *Q* output of the first or leftmost flip-flop set to 1 while the remaining three are 0.

Johnson Counter

The Johnson counter is similar to the ring counter except that the $\overline{Q}$ *output* of the last stage is fed back to the input of the first stage, as shown in Figure 5–48. Besides the term Johnson counter, this configuration is also referred to by other names, such as *twisted ring*, *walking ring*, and *switchtail* counters.

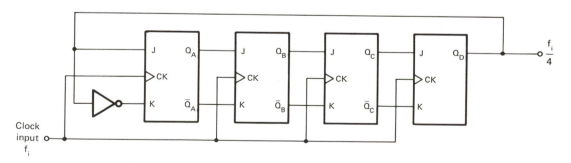

Fig. 5-46 Four-stage ring counter using *J–K* flip-flops.

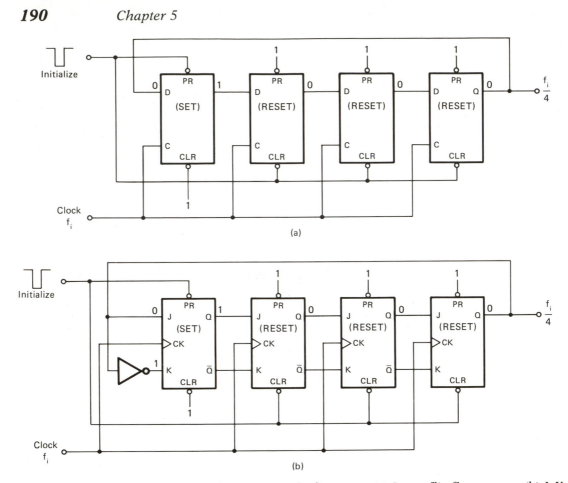

(a)

(b)

Fig. 5-47 Initializing a ring counter by presetting the first stage: (a) *D*-type flip-flop counter, (b) *J–K* flip-flop counter.

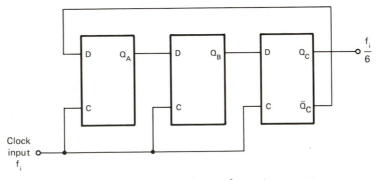

Fig. 5-48 Three-stage walking ring (Johnson) counter.

Although the Johnson and ring counters are similar except for the feedback connection, *the output frequency of the Johnson counter is one-half that of the ring counter.* Consequently, if a Johnson counter has *N* stages, the output frequency is

$$f_o = \frac{f_i}{2N} \qquad (5\text{--}4)$$

In addition, unlike the $1/N$ duty cycle associated with the ring counter, the output wave form of the Johnson counter is symmetrical, as shown in Figure 5–49. Like the ring counter, the *Q* output of each flip-flop is shifted by one input clock cycle with respect to each other.

As stated previously, the output frequency of a Johnson counter is one-half that of a ring counter. This implies that the Johnson counter requires only half the number of flip-flops for a given modulus that is required for a ring counter. For *even* modulos (2, 4, 6, etc.), no decoding gates are required for Johnson counters, but decoding gates are required for *odd-length* modulos.

To illustrate this point, Figure 5–50 shows how a 4018 CMOS walking ring decade counter can be configured for any modulus from 2 to 10. Note that the *even-length* configurations require only a direct connection from a particular stage output back to the counter's input pin. For *odd-length* counters, a 4081 CMOS 2-input AND gate is used to decode the output pair. Table 5–10 lists the available CMOS Johnson counters.

RATE MULTIPLIERS

The *rate multiplier* is a specialized counter circuit that produces an output pulse train whose frequency is proportional to the product of two inputs. As shown in Figure 5–51, the rate multiplier basically

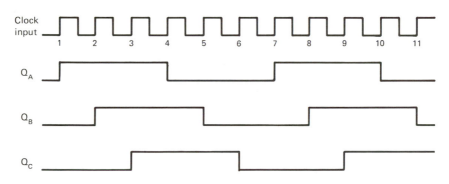

Fig. 5-49 Three-stage walking ring counter timing diagram.

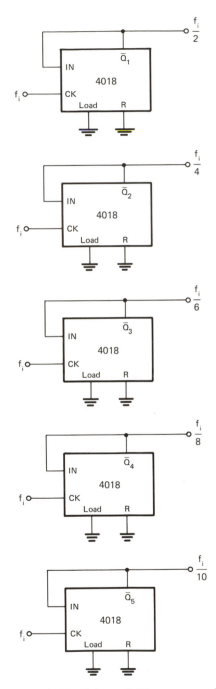

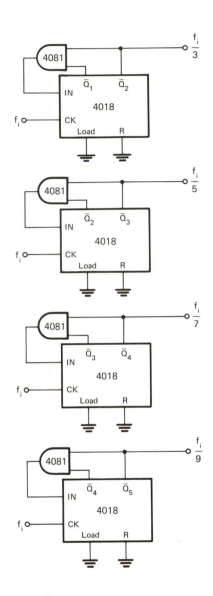

Fig. 5-50 Using a 4018 counter to divide an input frequency by any number from 2 to 10.

Table 5-10
Available CMOS Johnson (Walking Ring) Counters

Device	Type	Modulus	Remarks
4017	Decoded 1-of-10	10	Five-stage counter
4018	Programmable	2 to 10	—
4022	Decoded 1-of-8	8	Four-stage counter
4026	Decade	10	Has seven-segment display output
4033	Decade	10	Has seven-segment display output and ripple blanking I/O

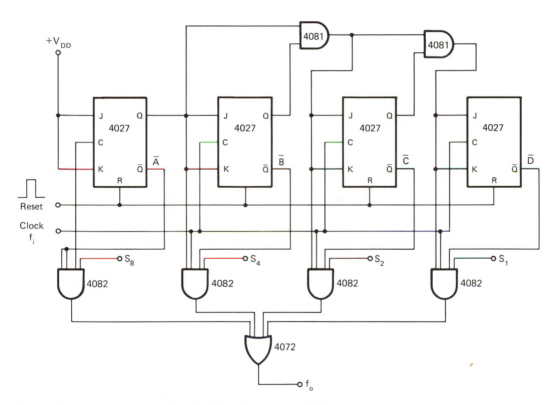

Fig. 5-51 Logic diagram of a CMOS 4-bit rate multiplier.

consists of four *J–K* flip-flops, two 2-input AND gates, four 4-input AND gates, and a single 4-input OR gate. One of the two required inputs is a clock frequency f_i, while the other is a preprogrammed 4-bit multiplier (S_1, S_2, S_4, and S_8), which can be either binary or BCD.

For a *binary* rate multiplier, the output frequency f_o will be $N/16$ times f_i, where N is a 4-bit binary number from 1 to 15. The *BCD* rate multiplier has an output frequency that ranges from one-tenth to nine-tenths times the input frequency. In either case, the output rate is always less than the input.

The timing diagram of Figure 5–52 shows how a value for $f_o = (6/16)f_i$ is obtained when the 2^2 (S_4) and 2^1 (S_2) programming inputs are 1 while the remaining two inputs are 0. From the timing diagram, the output of the rate multiplier consists of pulses that are *unevenly spaced*, although there are 6 output pulses for every 16 input cycles. Consequently, it must be recognized that the output rate is an *average rate*, so there is always a round-off error. However, this error can be reduced by increasing the bit capacity of the multiplier.

The 4089 and 4527 are, respectively, a CMOS *binary* rate multiplier and *BCD* rate multiplier, as shown by their pin diagrams in Figure 5–53. Each contains an internal synchronous 4-bit counter, which, together with the settings of the four BINARY or BCD RATE SELECT inputs, produces a specified output pulse train. The INHIBIT IN (INH), CASCADE (CAS), CLEAR (CLR), STROBE (STR), and SET-to-"9" (for the 4527) or SET-to-"15" (for the 4089) inputs are normally at logic 0.

As illustrated in Figure 5–54, two or more 4089 binary rate multipliers are cascaded, for which the output rate is given by the generalized relation

$$f_o = \frac{A \cdot B \cdot C \cdots N}{2^N}\,(f_i) \tag{5–5}$$

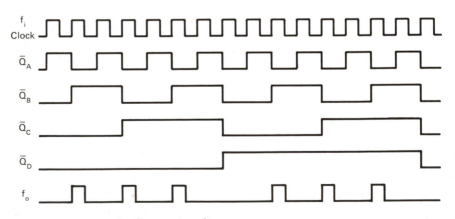

Fig. 5-52 Rate multiplier timing diagram.

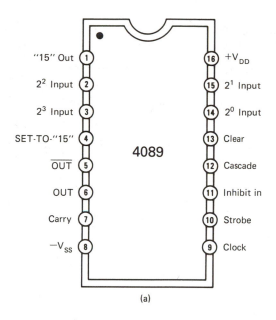

(a)

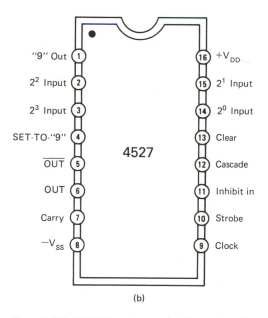

(b)

Fig. 5-53 CMOS rate multiplier pin diagrams: (a) 4089 binary rate multiplier, (b) 4527 BCD rate multiplier.

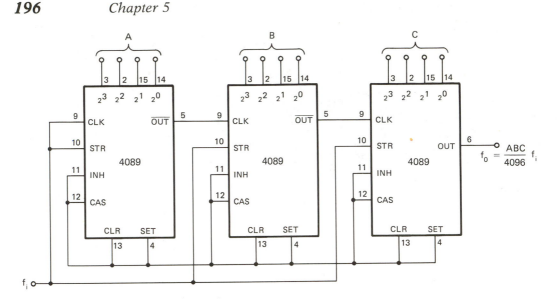

Fig. 5-54 Binary rate multipliers connected in series.

The numbers *A*, *B*, and so on, are the decimal equivalents of the 4-bit binary code for each section, and 2^N is the number of bits. If, for example, three 4089 devices are cascaded, then $2^N = 2^{(3)(4)}$, or 4096. The numbers *A*, *B*, and *C* individually would be any number between 1 and 15. Therefore, the maximum fraction of the input frequency in this case is $(15 \times 15 \times 15)/4096$, or $3375/4096$.

Using the 4527 BCD version, they are connected in parallel, as shown in Figure 5–55, forming a multibit multiplier. The output rate is

$$f_o = \frac{A}{10^N} (f_i) \qquad (5\text{--}6)$$

where *A* is the BCD number and *N* is the number of parallel stages. If, for example, two 4527 devices are connected in parallel, then *A* can range from 1 to 99, while the denominator of Eq. 5–6 equals 100. Consequently, the output rate can vary from 0.01 to 0.99 times the input rate f_i, depending on the settings of the programming inputs.

Multiplication

By using three or more rate multipliers and a single up/down counter (74190, 74191, 4510, 4516, etc.), we are able to obtain the product of two

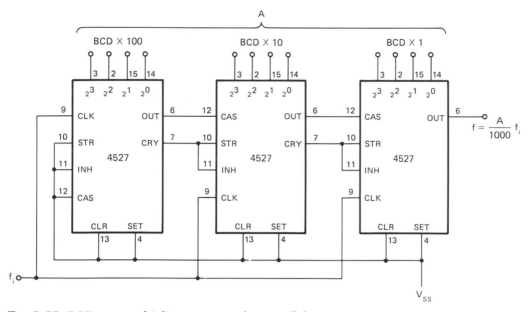

Fig. 5-55 BCD rate multipliers connected in parallel.

or more numbers. This product is not in terms of an output frequency, but rather a binary or BCD number that can be decoded to drive a display.

With the circuit of Figure 5–56, we can program two 4-bit variables, A and B, so that the 4-bit output of the up/down counter is

$$N = \frac{A \cdot B}{k} \tag{5–7}$$

where k is 10 when using a BCD rate multiplier and equal to 16 for a binary device. Using 4527 CMOS BCD rate multipliers and a 4510 CMOS BCD up/down counter as an example, we can better understand the operation of Figure 5–56. The output of the 4527 counter programmed with the 4-bit BCD number A is

$$f_o (A) = \frac{A}{10} f_{clock} \tag{5–8}$$

Similarly, the output of the 4527 device programmed with the 4-bit BCD number B is

$$f_o(B) = \frac{B}{10} f_o(A)$$

or

$$= \frac{AB}{100} f_{\text{clock}} \qquad\qquad (5\text{--}9)$$

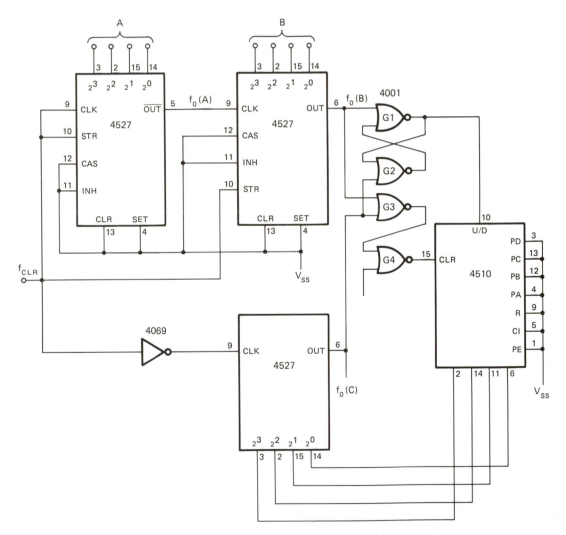

Fig. 5-56 Multiplication of two 4-bit numbers using rate multipliers.

It should be noted that, although these are *BCD* rate multipliers, they nevertheless are connected in *series*. The output rate of the remaining 4527 is

$$f_o\,(C) = \frac{N}{10}\,f_{\text{clock}} \qquad\qquad (5\text{--}10)$$

where the 4-bit BCD number N is determined by the output of the up/down counter (Eq. 5–7), or

$$f_o\,(C) = \frac{AB}{10}\,f_{\text{clock}} \qquad\qquad (5\text{--}11)$$

NOR gates *G*1 and *G*2, which are cross-coupled, convert the outputs of the *A* and *C* rate multipliers to a signal that tells the up/down counter whether to count up or down. The *B* output controls the up count, while the *C* output controls the down count. As these two outputs change, the up/down counter then counts up or down until the loop stabilizes, in which case

$$f_o\,(B) = \;f_o\,(C)$$

$$\frac{AB}{100}\;f_{\text{clock}} = \frac{N}{10}\,f_{\text{clock}}$$

from which we get $N = (AB)/10$.

Frequency Ratios

By using a pair of rate multipliers and a corresponding up/down counter, as shown in Figure 5–57, we are then able to compute the ratio of two input frequencies, f_A and f_B. Such a technique is the basis for several systems, such as the "miles per gallon" or MPG indicator used in automobiles.

Since the two input frequencies are usually asynchronous, care must be taken to make certain that the two inputs do not coincide. For rate multiplier *A*, whose program inputs are set at $DCBA = 0001$, or a decimal 1, the output rate is

$$f_o\,(A) = \frac{f_A}{10} \qquad\qquad (5\text{--}12)$$

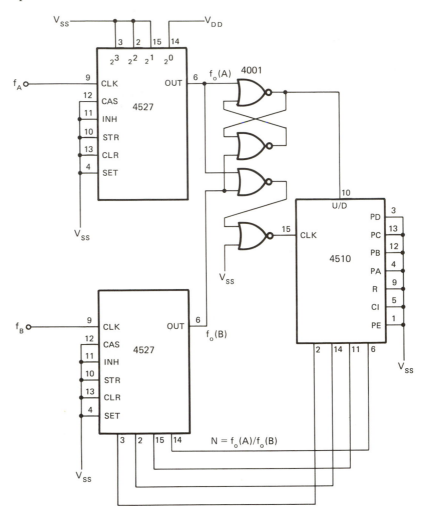

Fig. 5-57 Determining the ratio of two frequencies.

The output of rate multiplier *B*, whose program inputs are set by the 4-bit output *(N)* of the up/down counter, is

$$f_o\,(B) = \frac{Nf_B}{10} \qquad\qquad (5\text{--}13)$$

When the loop finally stabilizes, the two output rates are equal, so

$$f_o (A) = f_o (B)$$

$$\frac{f_A}{10} = \frac{N f_B}{10}$$

or

$$N = \frac{f_A}{f_B} \qquad\qquad (5\text{--}14)$$

Additional details concerning CMOS rate multipliers are fully described in the RCA Application Note, ICAN-6739, *COS/MOS Rate Multipliers— Versatile Circuits for Synthesizing Digital Functions*, by Gary J. Summers.

WAVEFORM GENERATION

Although operational amplifiers and other specialized integrated circuits are frequently used to generate periodic waveforms such as sine waves, triangles, and sawtooths, digital counters can also be used to generate periodic waveforms with respectable results.

The basis for the digital generation of sine waves was described by A. C. Davies[*] using what is termed a *transversal digital filter.* Using binary ripple and walking ring (Johnson) counters, different waveforms can be obtained. By properly adding the outputs of the counter, which are some binary function of the input with various delays, the principal harmonics of the input square wave clock signal can be filtered out. Consequently, only the fundamental component with several higher-order harmonics remains. The output is then a "stepped" or "chunky" approximation of a sine wave.

As shown in Figure 5–58, a multistage walking-ring counter is constructed from *D*-type flip-flops. For *N* flip-flop stages, the output frequency at the *Q* output of any flip-flop will be $1/(2N)$ times the input frequency. If five stages are used, then the output frequency will be one-tenth the input. As the clock frequency changes, so does the output. Using resistors, the outputs of *N* — 1 stages are then summed up to generate the sine wave, as shown for a four-stage counter in

[*]A. C. Davies, "Digital Generation of Low-Frequency Sinewaves," *IEEE Transactions on Instruments and Measurement*, Vol. IM-18, No. 2, June 1969, pp. 97–105.

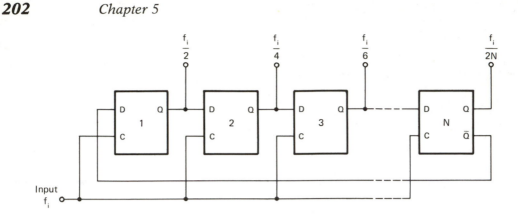

Fig. 5-58 Multistage walking ring counter.

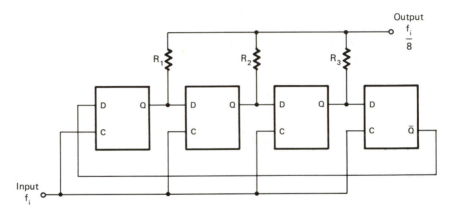

Fig. 5-59 Four-stage walking ring counter sine-wave generator.

Figure 5–59. Based on Davies's mathematics, the weighting resistors are determined by the generalized formula

$$R_N = \left[\frac{\dfrac{\sin(k\pi)}{N}}{\sin \dfrac{\pi}{N}} \right] R_I, \quad k = 1, 2, 3, \ldots (N-1) \qquad (5\text{--}15)$$

The lowest uncanceled harmonic is the $(2N - 1)$th, which has an amplitude of $1/(2N - 1)$ relative to the fundamental. For the four-stage sine-wave generator shown, the values of the resistors relative to R_1 are then

$$\frac{R_1}{R_1} = 1.000$$

$$\frac{R_2}{R_1} = 1.414$$

$$\frac{R_3}{R_1} = 1.000$$

The lowest uncanceled harmonic is (2) (4) − 1, or the seventh. If the clock frequency is, for example, 10 kHz, then the output sine wave will have a fundamental frequency of 10 kHz/8, or 1250 Hz, while the lowest uncanceled harmonic is the seventh, or a frequency of 8750 Hz. At 8750 Hz, the seventh harmonic amplitude is one-seventh of the fundamental. If a 27-kΩ resistor is used for R_1 and R_3, then a 39-kΩ 5% resistor is required for R_2.

Instead of using individual *D*-type flip-flops, walking ring CMOS counter devices, such as the 4018, can be used, as shown by the five-stage circuit of Figure 5–60a. However, as seen in Figure 5–60b,

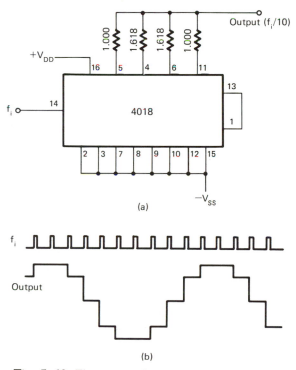

Fig. 5–60 Five-stage sine-wave generator using a 4018 counter: (a) circuit, (b) output wave form.

the resemblance to a sine wave is not very good. One possible solution is to increase the number of stages, thereby increasing the number of steps per cycle, but this solution has the disadvantage of requiring more resistors. Even then, the resulting waveform will still possess some rough edges.

Another approach is to add an active low-pass filter to the output of the generator, as shown in Figure 5–61. The filter's *cutoff frequency* should be well below the lowest uncanceled harmonic frequency.

Besides walking ring counters, binary ripple counters can be used to generate sine waves. As shown in Figure 5–62, the first three stages of a 4024 CMOS seven-stage ripple counter with a pair of 4070 XOR gates are used to create a four-stage generator. Note that only two resistors are required instead of three. An eight-stage circuit using a 4024 counter with a 4051 analog multiplexer is shown in Figure 5–63.

Using a programmable up/down counter, such as the 4029 in Figure 5–64, it is also possible to generate triangle waveforms with the help of a *R-2R ladder network*. As the counter is clocked up to its final

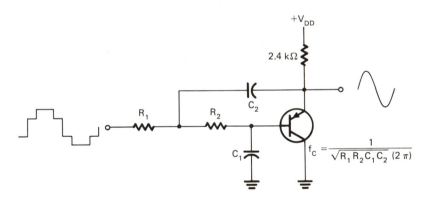

$$f_c = \frac{1}{\sqrt{R_1 R_2 C_1 C_2}\,(2\pi)}$$

(a) 2nd-order transistor circuit

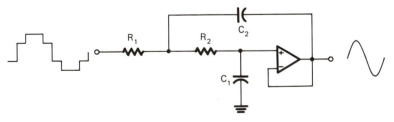

(b) 2nd-order op-amp circuit

Fig. 5-61 Active low-pass filter: (a) second-order transistor circuit, (b) second-order op-amp circuit.

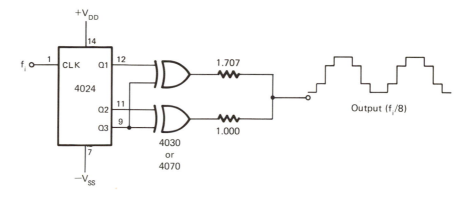

Fig. 5-62 Four-stage sine-wave generator using a 4024 CMOS binary ripple counter.

count, determined by the settings of its PRESET inputs, the CARRY OUT (*CO*) signal goes from 1 to 0, which is then inverted and used to clock the *D*-type flip-flop. The clocking action tells the 4029 whether to count up or down. When the down count reaches 0000, the *CO* signal

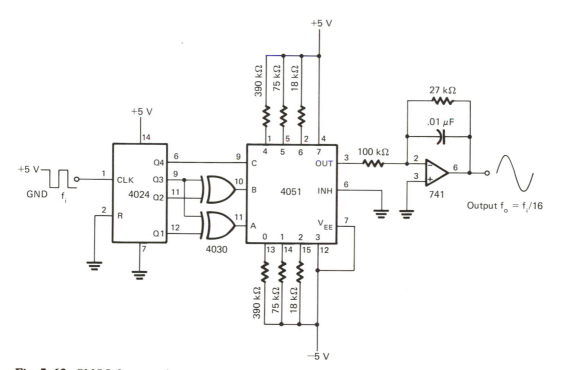

Fig. 5-63 CMOS 8-stage sine-wave generator.

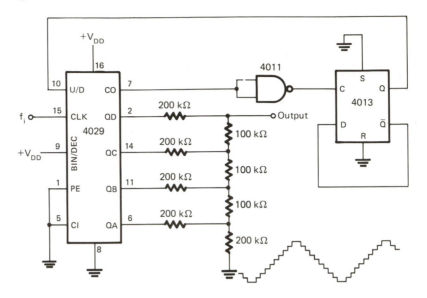

Fig. 5-64 CMOS triangle waveform generator.

clocks the flip-flop so that the counter counts up every other time the flip-flop is clocked.

As shown in Figure 5–65, a *R-2R* ladder network is used with a binary up counter, such as the 4520 with type 4016 CMOS transmission gates, to generate sawtooth waveforms in 16 equal voltage steps from ground to V_{DD}.

REVIEW

This chapter discussed how flip-flops are combined to create higher-order logic elements for the temporary storage and transfer of data: the shift register and the counter. We divided shift registers into two broad classes, depending on whether or not they accept serial or parallel data, and also on whether the shifted data are output in a serial or parallel format.

The latter half of this chapter discussed binary asynchronous (ripple) and synchronous counters and how they can be programmed, cascaded, or combined with other logic elements to synthesize common waveform shapes, as well as use of the rate multiplier to multiply and divide two frequencies.

INTRODUCTION TO THE EXPERIMENTS

The following experiments are designed to familiarize you with the operation and characteristics of several types of digital shift registers and counters. The experiments that you will perform can be summarized as follows:

Experiment	**Purpose**
5–1	Demonstrates the operation of a 74164 as both an 8-bit serial in–parallel out shift register and a serial in–serial out shift register.

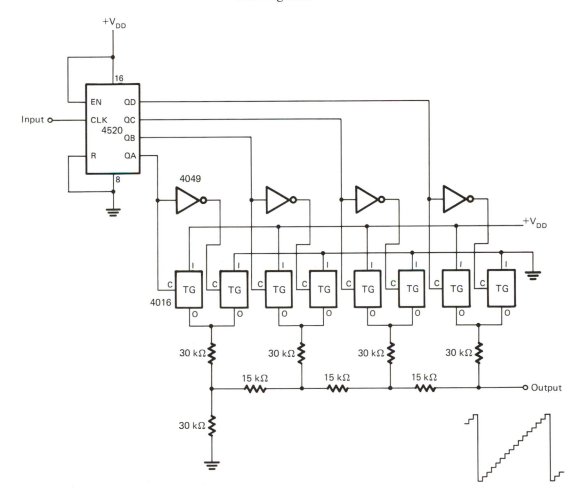

Fig. 5-65 CMOS sawtooth wave-form generator using transmission gates.

Experiment	Purpose
5–2	Demonstrates the operation of fixed modulus counters, using the 7490 TTL counter.
5–3	Demonstrates the operation of a programmable up/down counter using the 74192 TTL decade counter.
5–4	Demonstrates the operation of a sine-wave generator made from a 4018 CMOS five-stage Johnson counter.

EXPERIMENT 5–1

PURPOSE

Demonstrates the operation of a 74164 as both an 8-bit serial in–parallel out shift register and a serial in–serial out shift register.

PIN CONFIGURATION OF INTEGRATED-CIRCUIT DEVICE

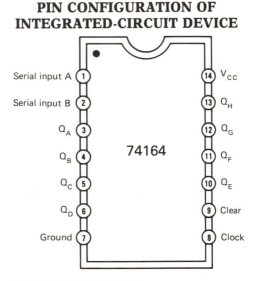

Fig. 5-66 74164 pin diagram.

Required Parts and Equipment

1 74164 TTL SIPO shift register
8 LED monitors
1 TTL logic switch
2 TTL logic pulsers
1 solderless breadboard
1 power supply (5 V)

SCHEMATIC DIAGRAM OF CIRCUIT

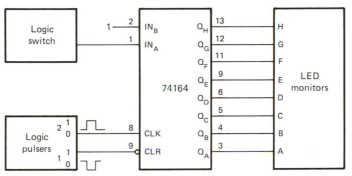

Fig. 5-67 Schematic diagram for Experiment 5-1.

STEP 1

Wire the circuit shown in the schematic diagram of Figure 5–67 and apply power to the breadboard.

STEP 2

Initially clear the register's contents by pressing and releasing pulser 1. Now set the logic switch, which represents the register's *serial* input data, at logic 1. Press and release pulser 2 eight times. After you release the pulser on the eighth time, all eight LED monitors should be lit.

Each time the shift register is clocked, going from logic 0 to 1, the serial input data to the first internal flip-flop stage was 1. As the register is repeatedly clocked, data move through the register.

STEP 3

Now set the logic switch at logic 0 and activate pulser 2 eight more times. All LED monitors should now be unlit.

STEP 4

To shift the 8-bit sequence 00000001 through the register, first clear the register using pulser 1 as in step 1. Next, set the logic switch at logic 1 and activate pulser 2 once. Then set the logic switch back to logic 0 and clock the shift register *eight* more times.

Each time you clock the register, the next succeeding LED monitor lights up while the LED monitor that was just lit is now unlit. Consequently, only one of the eight LED monitors is lit at any given time. The output whose LED monitor is lit is the *least significant* or *rightmost* bit of the binary serial input data sequence 00000001. Does the shift register shift from output Q_A toward Q_H or from Q_H to Q_A?

You should have observed that the serial input data *travel* from output Q_A toward Q_H, in which case the 74164 is *right shifting* the data. Since the outputs of all eight flip-flops are available *simultaneously*, the shift register is said to be a *parallel-out*-type register. However this same register can also act as a *serial-out* shift register if all but the last output (Q_H) are ignored.

EXPERIMENT 5-2

PURPOSE

The purpose of this experiment is to demonstrate the operation of fixed modulus counters, using the 7490 counter.

PIN CONFIGURATION OF
INTEGRATED-CIRCUIT DEVICE

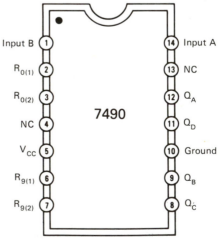

NC = No Connection

Fig. 5-68 7490 pin diagram.

Required Parts and Equipment

1   7490 TTL decade counter
1   square-wave clock (10 kHz)
1   solderless breadboard
1   frequency counter
1   power supply (5 V)

SCHEMATIC DIAGRAM OF CIRCUIT

STEP 1

Wire the circuit shown in the schematic diagram of Figure 5–69 and then apply power to the breadboard.

STEP 2

With a frequency counter, first adjust the input clock frequency (input *A*, pin 14) as close as possible to 10 kHz, and then measure the output frequency at pin 12.

You should measure an output frequency that is *one-half* the input, or 5 kHz (within 1 Hz). For this part, the 7490 is wired as a *divide-by-2* counter, so the modulus is 2.

STEP 3

Next, connect the 10-kHz clock to pin 1 (input *B*) of the 7490 counter and connect the frequency counter's input to pin 11 as shown in Figure 5–70. For a 10-kHz input, what is the output frequency?

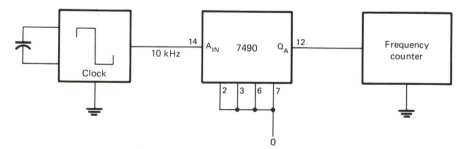

Fig. 5–69 Schematic diagram for Experiment 5-2.

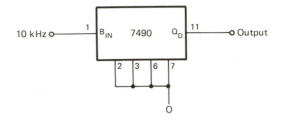

Fig. 5-70 Schematic diagram for Experiment 5-1, step 3.

You should measure an output frequency that is exactly *one-fifth* the input frequency, or 2 kHz (within 1 Hz). For this part, the 7490 is wired as a *divide-by-5* counter, so the modulus is 5.

STEP 4

Wire the 7490 counter as shown in Figure 5–71. It should now be apparent that the 7490 internally has both a divide-by-2 and a divide-by-5 counter section. This new arrangement has the output of the divide-by-2 section connected to the input of the divide-by-5 section. For a 10-kHz input, what is the measured output frequency at pin 11?

You should now measure an output frequency that is exactly *one-tenth* the input, or 1 kHz. Both internal counter sections are cascaded together to give a total modulus of 10. This follows the rule that when two or more counters are *cascaded,* or connected in series, the modulus of the entire circuit is the *product* of the moduli of the individual gates, so that, for this experiment, 5×2 gives an overall modulus of 10. Also, as a general rule, when

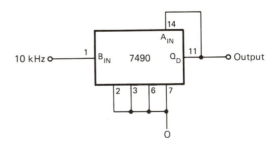

Fig. 5-71 Schematic diagram for Experiment 5-1, step 4.

counters are cascaded, they are arranged in decreasing order of their maximum rated input frequency. The "fastest" counter is placed first, and so on.

EXPERIMENT 5-3

PURPOSE

The purpose of this experiment is to demonstrate the operation of a programmable up/down counter using the 74192 TTL decade counter.

PIN CONFIGURATION OF INTEGRATED-CIRCUIT DEVICE

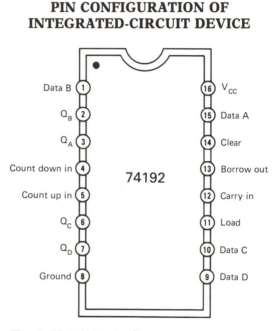

Fig. 5-72 74192 pin diagram.

Required Parts and Equipment

1   74192 TTL programmable up/down decade counter
2   TTL logic pulsers
4   TTL logic switches
2   TTL LED monitors
1   seven-segment LED display with decoder/driver module

1   square-wave clock (10 kHz)
1   solderless breadboard
1   frequency counter
1   power supply (5 V)

SCHEMATIC DIAGRAM OF CIRCUIT

STEP 1

Wire the circuit shown in the schematic diagram of Figure 5–73, initially presetting all four logic switches at logic 0. Both pulsers have their outputs normally at logic 1. If you do not have a seven-segment LED display module, use the circuit of Figure 7–13.

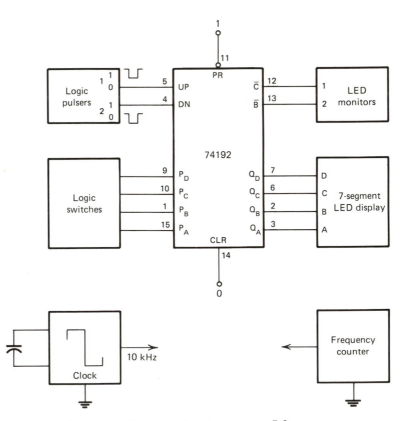

Fig. 5-73 Schematic diagram for Experiment 5-3.

STEP 2

Apply power to the breadboard. What number appears on the seven-segment LED display?

Depending on the initial state of the internal logic of the 74192 counter, any number between 0 and 9 could appear when power is first applied. When this experiment was performed on one occasion, the number 0 was displayed. In addition, both LED monitors were lit. However, your results may differ.

STEP 3

Press and release pulser 1, which is connected to the counter's UP input at pin 5, several times. What do you see happening to the seven-segment LED display as you press and release pulser 1 several times?

You should notice that the LED display indicates that the counter counts *upward:* 0, 1, 2, 3, etc.

STEP 4

Continue to press and release pulser 1. What happens to the 7-segment LED display *and* LED monitor 1 after the displayed count goes past 9?

After the 74192 decade counter reaches its maximum count of 9, the CARRY signal at pin 12, indicated by LED monitor 1, briefly goes out, and when the numerical display reads 0, LED monitor 1 lights up again. When several 74192 counters are cascaded to form multidecade counters, the CARRY output signal is used as the count-up input for the next higher decade.

STEP 5

Continue to activate pulser 1 until the number 6 appears on the LED display. Now press and release pulser 2, which is connected to the DN input of the counter (pin 4), three times. What happens to the LED display?

You should see that the display counts *downward:* 6, 5, 4, 3.

STEP 6

Activate pulser 2 five more times. What happens to both the seven-segment LED display and LED monitor 2 after the displayed count goes past 0?

> After the counter reaches its minimum count of 0, the BORROW output signal (LED monitor 1) goes out briefly and then lights up again when the display reads 9. When several 74192 counters are cascaded to form multidecade counters, the BORROW output signal is used as the count-down input for the next higher decade.

STEP 7

Remove the power from the breadboard. Now connect pin 11 of the 74192 counter to pin 13 instead of the +5-V supply. In addition, disconnect both LED monitors from the counter.

STEP 8

Set the PRESET logic switches so that $DCBA = 0111$ and apply power to the breadboard. What number do you see on the digital display?

> Again, depending on the initial state of the internal logic of the 74192 counter, any number between 0 and 9 may appear when power is first applied. When this step was performed on one occasion, the number 4 was displayed. However, your results may differ.

STEP 9

Now activate pulser 2 until the digital display reads 0. Press and release the pulser again. What number now appears on the display?

> When the down pulser is pressed, the digital display should briefly read 7, and then read 6 when the pulser is released. When the displayed count down goes past zero, the 74192 counter is *parallel loaded* with the 4-bit BCD number that is set by the logic switches connected to the PRESET inputs. In this step, the BCD code is 0111, which is the number 7. When the pulser is released, the LOAD input is now at logic 1; the number 6 is displayed.

Consequently it now takes *seven* input clock cycles (6, 5, 4, 3, 2, 1, and 0) until the counter is again loaded with the 4-bit BCD number at the PRESET inputs.

STEP 10

Now change the PRESET input logic switches to some other 4-bit BCD number between 0010 (2) and 1010 (10). Convince yourself that each time the displayed count down passes 0, when you release the pulser, the displayed number is one less than the equivalent BCD code set by the logic switches. The number set by the logic switches is the number of clock pulses needed to make the display pass zero again.

If you change the BCD code at the PRESET inputs during the counting process, the new number will not be loaded until the count passes zero.

STEP 11

Disconnect the power from the breadboard and connect a 10-kHz square-wave clock signal to pin 4 of the 74192 counter. Apply power to the breadboard and set the PRESET logic switches so that $DCBA = 0001$.

STEP 12

With a frequency counter, first adjust the input clock frequency (pin 4) as close as possible to 10 kHz, and then measure the output frequency at pin 13.

You should measure the same value for both the input and output frequencies, 10 kHz. Since the PRESET logic switches are set at $DCBA = 0001$, or the decimal number 1, the 74192 has now been parallel loaded with the number 1, so we now have a modulus-1 (i.e., $N = 1$) counter. Consequently, the output frequency is the input frequency divided by the modulus N. In this case, both are the same 10-kHz frequency. For a single counter stage, it really does not make too much sense to have a modulus-1 counter.

STEP 13

Now vary the PRESET logic switches according to Table 5–11, and record the measured output frequency for each setting.

Table 5-11
Data for Experiment 5-3

| PRESET Data | | | | Measured Output |
D	C	B	A	Frequency
0	0	0	1	10 kHz
0	0	1	0	
0	0	1	1	
0	1	0	0	
0	1	0	1	
0	1	1	0	
0	1	1	1	
1	0	0	0	
1	0	0	1	
1	0	1	0	
1	0	1	1	
1	1	0	0	
1	1	0	1	
1	1	1	0	
1	1	1	1	

From your measurements, you should have observed that the output frequency is that fraction of the 10-kHz input set by the decimal equivalent of the 4-bit BCD code. You have now used a single 74192 counter to divide the 10-kHz input frequency by any number from 1 to 15.

STEP 14

What is the output frequency when you set the PRESET logic switches at $DCBA = 0000$?

Since division by 0 is not defined in a strict mathematical sense, the 74192 internally becomes disabled so that the output frequency is the same as the input (10 kHz). Therefore, if the BCD PRESET inputs are set either to 0 or 1, the output is the same as the input.

EXPERIMENT 5-4

PURPOSE

The purpose of this experiment is to demonstrate the operation of a sine-wave generator made from a 4018 CMOS five-stage Johnson counter.

PIN CONFIGURATION OF INTEGRATED-CIRCUIT DEVICE

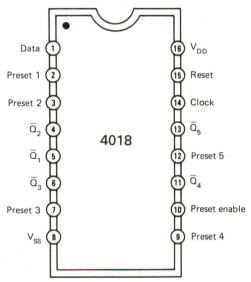

Data	1	16	V_{DD}
Preset 1	2	15	Reset
Preset 2	3	14	Clock
$\bar{Q}_2$	4	13	$\bar{Q}_5$
$\bar{Q}_1$	5	12	Preset 5
$\bar{Q}_3$	6	11	$\bar{Q}_4$
Preset 3	7	10	Preset enable
V_{SS}	8	9	Preset 4

Fig. 5-74 4018 pin diagram.

Required Parts and Equipment

- 2 24-kΩ resistors
- 2 39-kΩ resistors
- 2 56-kΩ resistors
- 2 0.082-μF capacitors
- 1 4018 CMOS five-stage Johnson counter
- 1 square-wave clock (approximately 1 kHz)
- 1 solderless breadboard
- 1 dual-trace oscilloscope
- 1 frequency counter
- 1 power supply (5 V)

SCHEMATIC DIAGRAM OF CIRCUIT

STEP 1

Wire the circuit shown in the schematic diagram of Figure 5–75. Next, set your oscilloscope to the following initial settings:

- Channel 1: 5 V/division
- Channel 2: 2 V/division
- Time base: 1 ms/division
- Triggering: channel 2, dc coupling

STEP 2

Apply power to the breadboard. Then position the input waveform (channel 1) on the oscilloscope's display so that it is above the output waveform of channel 2. Adjust the input square wave so that there are 10 complete cycles for the 10 horizontal scale divisions. The input frequency is then 1 kHz.

You should now notice that the output waveform, although a "chunky" sine wave, has a frequency that is one-tenth the input, or 100

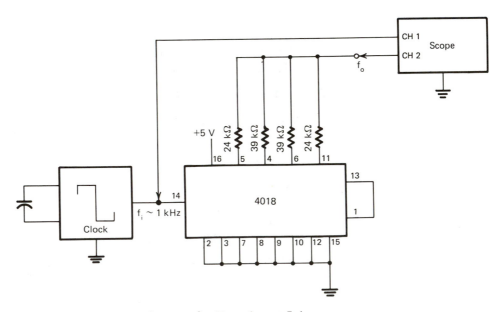

Fig. 5-75 Schematic diagram for Experiment 5-4.

Hz. This is because the 4018 is a five-stage Johnson counter. For an
N-stage counter, the output frequency will be $1/(2N)$ times the input.

STEP 3

Now change the oscilloscope's time base to 2 ms/division in order
to display two full cycles of the output waveform. Your output waveform should look like that shown in Figure 5–76.

STEP 4

Disconnect the power from the breadboard and add the simple,
passive low-pass filter shown in Figure 5–77 to the output of the 4018
sine-wave generator circuit. Now connect the channel 1 probe of the
oscilloscope to the input of the filter and the channel 2 probe to the
filter's output. How does the filter's output waveform compare with
the filter's input signal?

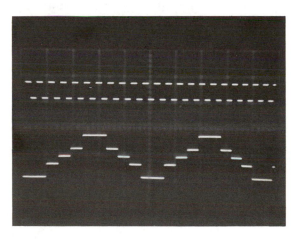

Fig. 5–76 ˙Oscilloscope display for step 3.

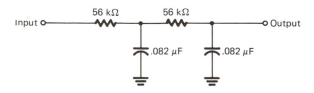

Fig. 5–77 Passive low-pass filter for Experiment 5-4,
step 4.

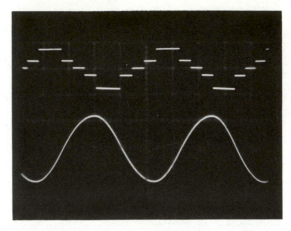

Fig. 5-78 *Oscilloscope display for step 4.

The output waveform should resemble a cleaner looking sine wave than the input signal. The low-pass filter removes rough edges of the waveform attributed to the higher harmonic components of the "chunky" waveform, leaving essentially only the fundamental 100-Hz sine wave, as shown in Figure 5–78.

CODE CONVERTERS, MULTIPLEXERS, AND DEMULTIPLEXERS

INTRODUCTION

Unlike the digital counters in the previous chapter, which relied primarily on flip-flops and a few logic gates for their function, this chapter discusses those functions that are basically constructed of gates. Like the combinatorial logic circuits discussed in Chapter 2, logic gates can be combined to produce decoders, encoders, multiplexers, and demultiplexers.

OBJECTIVES

After completing this chapter, you will be able to understand the operation of:

- *Binary/decimal code converters*
- *Binary/octal code converters*
- *Priority encoders*
- *Multiplexers*
- *Demultiplexers*

223

CODE CONVERTERS

A *code converter* is a device that translates one type of code into another. In terms of digital logic, an example of such a code converter would be a circuit that tells us the decimal equivalent of a 4-bit BCD number. Traditionally, code converters are divided into *decoders* and *encoders*. The preceding example is usually cited as an example of a *decoder*, in that this type of circuit converts the "unfamiliar" BCD number into a more "familiar" decimal number form. On the other hand, *encoders* have the opposite function.

Sometimes there is a fine line between what code is familiar and what is unfamiliar. Perhaps an analogy can be made by considering the following worthwhile question: "Is the beer glass half empty, or is it half full?" The reply depends on whether you are an optimist or a pessimist. What one calls a decoder or an encoder really depends on what side of the fence you're on, and some people have used these terms interchangeably. In this chapter, no hard and fast rules will be made as to what are decoders and encoders, as there will sometimes be exceptions to the rules. In short, both are *code converters*. In discussing specific circuits, the commonly used term, whether it be decoder or encoder, will be placed in parentheses. However, it will be left to you, the reader, to decide what *you* wish to call it.

BCD/Decimal Converters

Figure 6–1 shows the required logic for a BCD-to-decimal code converter (decoder), in that there is one output of the ten that is different from the other nine for a given 4-bit BCD input. Table 6–1 gives the truth table for such a circuit. Because there are four inputs and ten outputs, this type of code converter is frequently referred to as either a *4 line-to-10 line* or *1-of-10* decoder.

The 7442 TTL device has the equivalent function of Figure 6–1, while the 4028 and 74C42 are the equivalent CMOS functional devices, as shown by the pin diagrams of Figure 6–2. Using any of these devices, this 4 line-to-10 line arrangement can also be wired to function as a 2 line-to-4 line (1-of-4) converter. By permanently setting both the *D* and *C* inputs at 0, Figure 6–3 shows the 7442 wired as a 2 line-to-4 line decoder. Consequently, the maximum binary code is 0011, which is the decimal number 3, and results in the four possible decimal outputs of 0

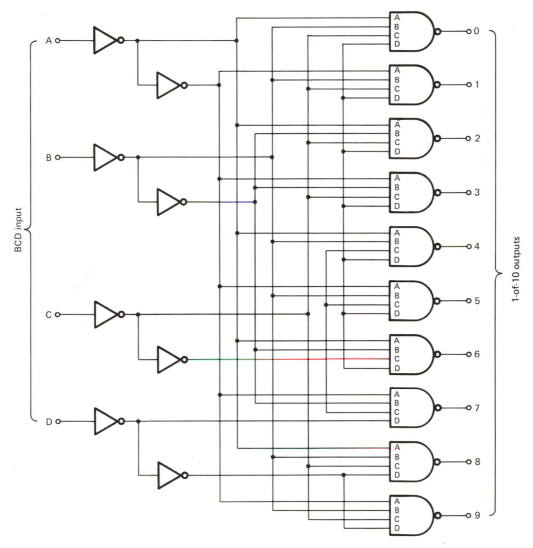

Fig. 6-1 BCD-to-decimal code converter.

through 3. Chapter 7 discusses the 7445 1-of-10 decoder for digital displays using incandescent lamps.

On the other hand, as shown by the decimal-to-BCD converter (encoder) in Figure 6–4, we can go the opposite way to convert one of ten possible inputs to its equivalent 4-bit BCD code. A logic 1 only on one of the ten inputs produces the equivalent BCD code.

Table 6-1
Truth Table for a BCD-to-Decimal Code Converter

BCD Input				Output									
D	C	B	A	0	1	2	3	4	5	6	7	8	9
0	0	0	0	0	1	1	1	1	1	1	1	1	1
0	0	0	1	1	0	1	1	1	1	1	1	1	1
0	0	1	0	1	1	0	1	1	1	1	1	1	1
0	0	1	1	1	1	1	0	1	1	1	1	1	1
0	1	0	0	1	1	1	1	0	1	1	1	1	1
0	1	0	1	1	1	1	1	1	0	1	1	1	1
0	1	1	0	1	1	1	1	1	1	0	1	1	1
0	1	1	1	1	1	1	1	1	1	1	0	1	1
1	0	0	0	1	1	1	1	1	1	1	1	0	1
1	0	0	1	1	1	1	1	1	1	1	1	1	0

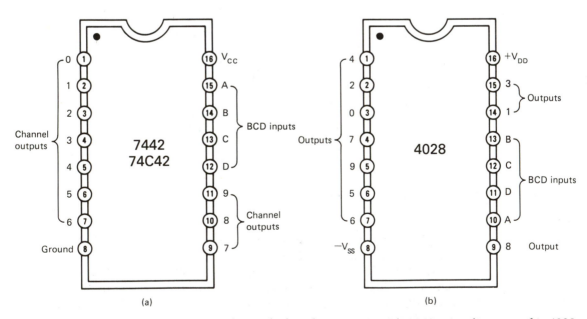

Fig. 6-2 Integrated-circuit BCD-to-decimal decoders: (a) 7442/74C42 pin diagram, (b) 4028 pin diagram.

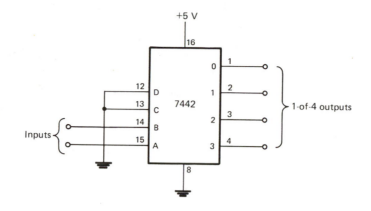

Fig. 6-3 Two-to-four line decoder using a 7442.

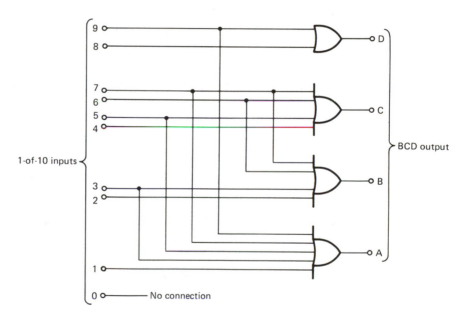

Fig. 6-4 Decimal-to-BCD code converter using OR gates.

Binary/Octal Converters

Figure 6–5 shows the logic for a binary-to-octal, *3 line-to-8 line converter* (decoder); its associated truth table is given in Table 6–2. Although there are no specific TTL and CMOS devices that accomplish

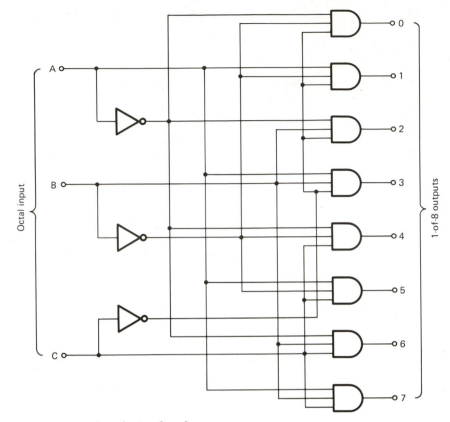

Fig. 6-5 Octal-to-decimal code converter.

Table 6-2
Truth Table for a Binary-to-Octal Code Converter

Binary Input			Output							
C	B	A	0	1	2	3	4	5	6	7
0	0	0	0	1	1	1	1	1	1	1
0	0	1	1	0	1	1	1	1	1	1
0	1	0	1	1	0	1	1	1	1	1
0	1	1	1	1	1	0	1	1	1	1
1	0	0	1	1	1	1	0	1	1	1
1	0	1	1	1	1	1	1	0	1	1
1	1	0	1	1	1	1	1	1	0	1
1	1	1	1	1	1	1	1	1	1	0

228

this conversion directly, we can nevertheless use a BCD-to-decimal converter (7442, 74C42, or 4028) with its *D* input permanently wired at logic 0 to achieve the same effect, as shown in Figure 6–6. Consequently, the maximum binary code possible is 0111, which is the decimal number 7, and results in the eight possible decimal outputs of 0 through 7.

Going the opposite way, Figure 6–7 shows the required logic for an octal-to-binary, or 8 line-to-3 line converter (encoder). Only *one* of the

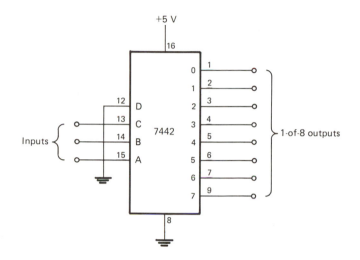

Fig. 6-6 Octal-to-decimal code converter using a 7442.

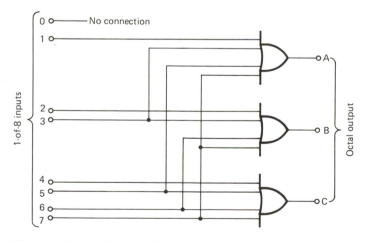

Fig. 6-7 Decimal-to-octal code converter using OR gates.

eight input lines is at logic 1 at any given time; otherwise, the 3-bit binary output would be in error.

Priority Encoders

With the encoders previously discussed, there was the condition that only one input could be at logic 1 at any given time; otherwise, the resultant encoder output would be in error. A variation of the previously discussed encoders is that *multiple logic inputs are allowed*. However, a *priority* is assigned so that the encoder produces a BCD output that equals the *highest decimal number* that is present on the inputs; hence the name, *priority encoder*. As an example, if a priority 8 line-to-3 line encoder has logic 1s on the 5 and 2 inputs, then the BCD output will be 101, representing the decimal number 5, since it is the highest decimal number at logic 1. Consequently, the 7 input has the highest priority; the 0 input has the lowest. Table 6–3 lists the available TTL and CMOS priority encoders.

Other Code Converters

In addition to the code converters discussed up to this point, there is one remaining major class whose use is primarily associated with digital displays. Instead of discussing those converters here, the next chapter on digital displays specifically focuses on these types.

MULTIPLEXERS

A *multiplexer* or *data selector* is a device that has the ability to select one of a number of inputs and pass this input signal onto a single output path. This operation is analogous to a single-pole, multiposition switch,

Table 6-3
Available TTL and CMOS
Priority Encoders

Function	TTL	CMOS
8 line to 3 line	74148	4532
10 line to 4 line	74147	74C147

as shown in Figure 6–8. Depending on the position of the switch's pole, or wiper, we have the ability to select one from a possible four inputs, or *channels*. Once selected, the output signal is the same as the selected input.

In this example, we have to *manually* position the wiper to select the proper input channel. Now consider the logic diagram for the 4-input multiplexer of Figure 6–9a. In addition to four data channels, there are two *data select* input lines, which use a 2-bit binary code to select one of the four inputs. If a binary 01 ($S1 = 0$, $S0 = 1$), which is equal to the decimal number 1, is applied to the data select inputs, then the signal at input channel $D1$ will appear at the output, as summarized in Table 6–4.

Table 6–5 lists the available TTL and CMOS multiplexers. The 74157/74C157 and 74158/74C158 each have four 2-input multiplexers in a single package controlled by a common select input. The 74157 has a noninverted output, while the 74158 is inverted. Both the 74153/74C153 and 74253/74C253 are dual 4-input multiplexers. The only difference is that the 74153 has two *independent* pairs of select inputs, one pair for each section, while the select input for the 74253 controls both sections simultaneously, in addition to having three-state outputs. The 74151 and 74152 are the same except that only the 74151 provides complementary outputs. The 74150/74C150 provides only an inverted output.

All multiplexers listed in Table 6–5 have a *strobe*, or *enable* input, except the 74152. Generally, a logic 0 at this input is used to *strobe* or *enable* the multiplexer to function; otherwise, a 1 inhibits operation, and a specified logic level, depending on the device, appears as the output. As an example, Figure 6–10a is the same 4-input multiplexer as Figure 6–9a except for the added strobe/enable (STB) capability. If STB = 1, then output $Q = 0$ for *all* select input combinations. Only when STB = 0 is the output controlled by the select inputs. Thus the strobe input allows us to select one multiplexer out of several independent ones. An alternative name for the strobe input is the *chip select* input. The truth table for the 74153 is given in Table 6–6.

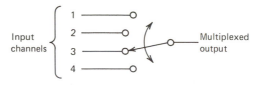

Fig. 6-8 Four-channel multiplexer using a SP4T switch.

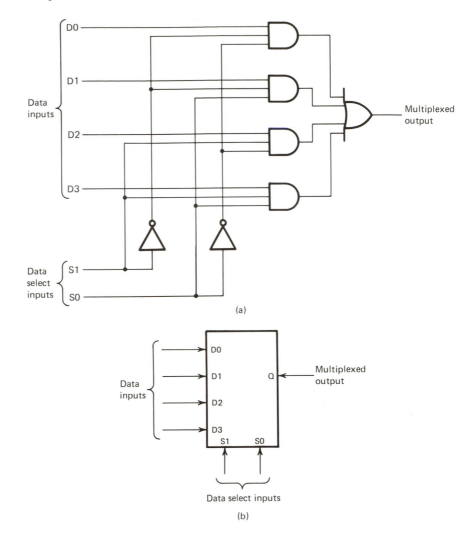

Fig. 6-9 Four-channel multiplexer: (a) logic diagram, (b) schematic symbol.

DEMULTIPLEXERS

A *demultiplexer* or *data distributor* has the opposite function of a multiplexer. It is a device that has the ability to direct a single signal to one of a number of outputs. This operation is analogous to a single-pole, multiposition switch, as shown in Figure 6–11. Depending on the

Table 6-4
Truth Table for a 4-Input Multiplexer

Data Select Input		Selected Output
S1	*S0*	Channel
0	0	*D0*
0	1	*D1*
1	0	*D2*
1	1	*D3*

Table 6-5
Available TTL and CMOS Multiplexers

Input Channels	Select Inputs	TTL	CMOS
2 (4)	1	74157, 74158, 74257[a]	74C157, 74C158, 74C257[a]
4 (2)	2	74153, 74253[a]	74C153, 74C253[a]
8	3	74151, 74152, 74251[b]	74C151, 74C251[b]
16	4	74150	74C150, 82C19[a]

[a]Three-state output.
[b]Complementary three-state output.

Table 6-6
Truth Table for the 74153
4-Input Multiplexer

Data Select Input		Strobe	Output
S1	*S0*	STB	*Q*
×	×	1	0
0	0	0	*D0*
0	1	0	*D1*
1	0	0	*D2*
1	1	0	*D3*

× = don't care.

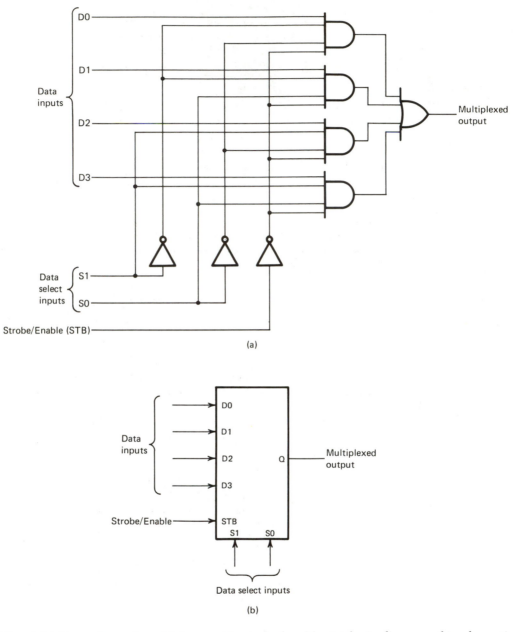

(a)

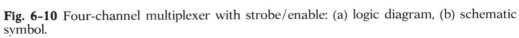

(b)

Fig. 6-10 Four-channel multiplexer with strobe/enable: (a) logic diagram, (b) schematic symbol.

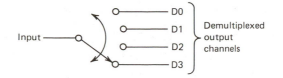

Fig. 6-11 Four-channel demultiplexer using a SP4T switch.

position of the switch's wiper, we have the ability to select one from a possible four output channels. Once selected, the output signal is the same as the input.

In this example, we have to *manually* position the wiper to select the proper input signal. Now consider the logic diagram for the four-line demultiplexer of Figure 6–12a. In addition to four output channels, or lines, there are two *data select* input lines, which use a 2-bit binary

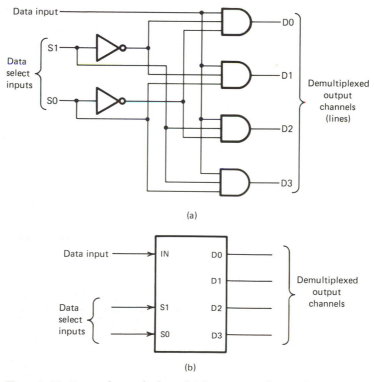

Fig. 6-12 Four-channel demultiplexer: (a) logic diagram, (b) schematic symbol.

code to select one of the four outputs, as summarized by the truth table of Table 6–7.

Table 6–8 lists the commonly available TTL and CMOS demultiplexers. All the devices listed, in addition to being demultiplexers, also serve as *decoders*.

As an example, the 74154, whose pin diagram is shown in Figure 6–13, has a data input (*G*1, pin 18) and an enable input (*G*2, pin 19), in addition to the four channel select inputs (*S*0, *S*1, *S*2, and *S*3). As illustrated in Figure 6–14, the 74154 functions as a *demultiplexer* by placing a logic 0 on the enable input while the input signal is connected to the data input. This input signal is then passed onto one of the 16 possible outputs determined by the 4-bit hexadecimal code at the channel select inputs. As shown, if *S*3 = 1, *S*2 = 0, *S*1 = 1, and *S*0 = 1, which is the equivalent of the decimal number 11, then the input signal present at *G*1 also appears at the output pin of channel 11 (pin 13), as the remaining 15 outputs are all logic 1. Setting the enable input at logic 1 disables the demultiplexer function so that all 16 channels are at logic 1.

On the other hand, if both *G*1 and *G*2 are set at logic 0, the 74154 functions as a simple *1-of-16 decoder*, as shown in Figure 6–15. As

Table 6–7
Truth Table for a 4-Input Multiplexer

| Data Select Input | | Selected Output |
*S*1	*S*0	Channels
0	0	*D*0
0	1	*D*1
1	0	*D*2
1	1	*D*3

Table 6–8
Commonly Available TTL and CMOS Demultiplexers

Output Channels	Select Inputs	TTL	CMOS
4 (2)	2	74139	74C139
8	3	74138	74C138
16	4	74154	74C154

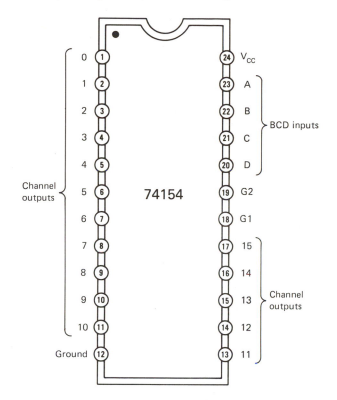

Fig. 6-13 74154 TTL 16-channel demultiplexer pin diagram.

shown, if $S3 = 0$, $S2 = 1$, $S1 = 1$, and $S0 = 0$, a logic 0 appears at channel 6 output, while the remaining 15 outputs have a logic 1.

As we have seen, the 74154 is a very handy device. Although it can only handle a maximum of 16 channels using a single package, we nevertheless are able to expand the number of decoded or multiplexed channels to any desired number. Figure 6–16 illustrates how two 74154s are connected to form a 32-state decoder/demultiplexer. As shown, this arrangement illustrates the demultiplexer mode, as the input data signal is applied simultaneously to both the $G1$ inputs. If the $G1$ inputs of both stages are tied at logic 0, then the circuit functions as a decoder.

Figure 6–17 shows how four 74154s are connected to create a 1-of-64 decoder. Five data select inputs are required to fully decode all 64 possible states.

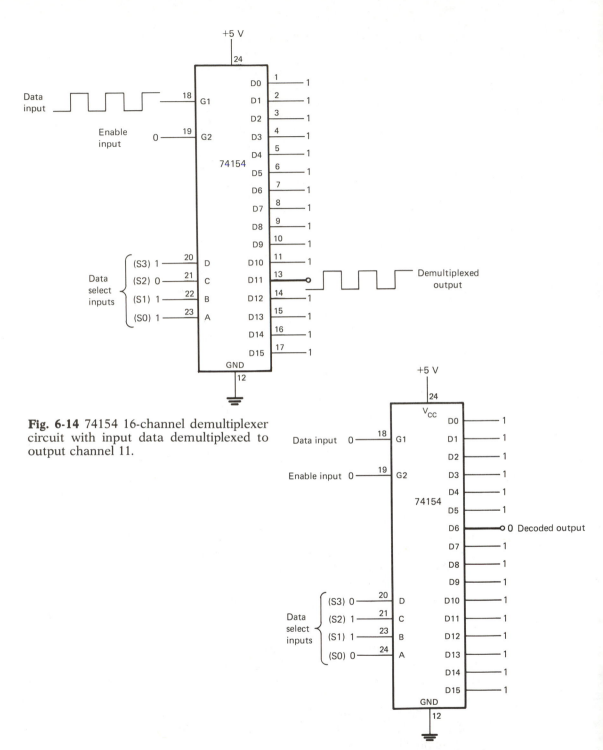

Fig. 6-14 74154 16-channel demultiplexer circuit with input data demultiplexed to output channel 11.

Fig. 6-15 74154 4-to-16 line decoder.

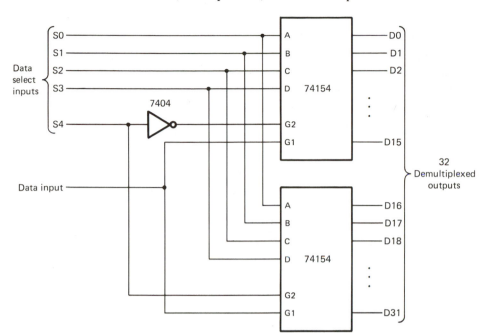

Fig. 6-16 Thirty-two channel demultiplexer.

CMOS MULTIPLEXER/DEMULTIPLEXERS

In the CMOS family of integrated-circuit devices, several devices have the capability of functioning both as a multiplexer and a demultiplexer, depending on what is considered input and output. As shown by the block diagram of Figure 6-18, the circuit is basically composed of a series of transmission gates with some type of BCD or hexadecimal decoder. Since CMOS bidirectional transmission gates are used, this arrangement allows the multiplexing or demultiplexing of analog as well as digital signals. Consequently, data sheets often refer to this device as an *analog multiplexer/demultiplexer.* Table 6-9 lists the commonly available CMOS devices, while the corresponding pin diagrams are given in Figure 6-19.

For the 4051, 4052, and 4053 devices, the transmission gate ON resistance is typically 150 Ω, whereas it is 200 Ω for the 4067 and 4097.

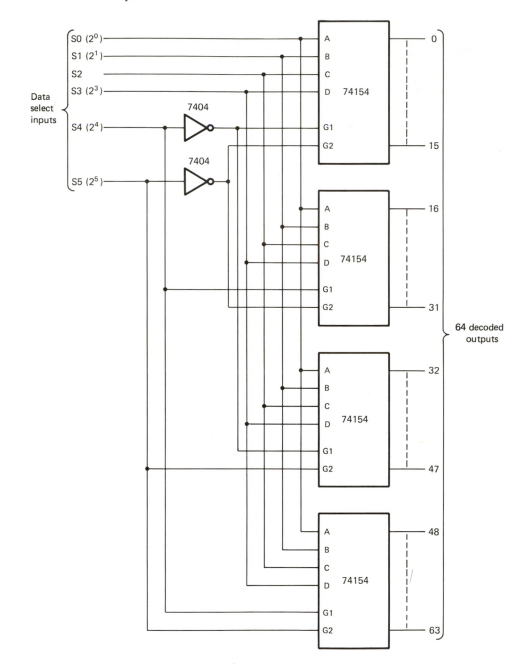

Fig. 6-17 Sixty-four channel decoder.

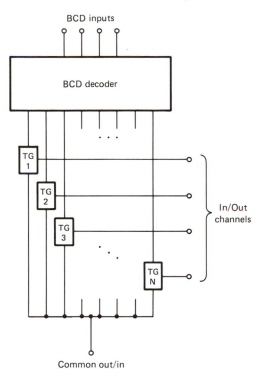

Fig. 6-18 Block diagram for a CMOS multiplexer/demultiplexer.

Table 6-9
Available CMOS Multiplexers/Demultiplexers

Device	Sections	Channels
4053	3	2
4052	1	4[a]
4051	1	8
4097	1	8[a]
4067	1	16

[a]Differential channel pairs.

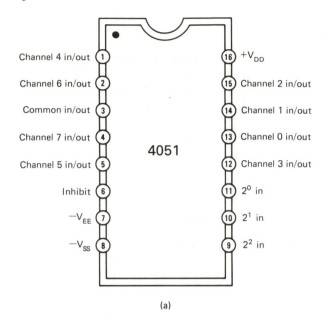

(a)

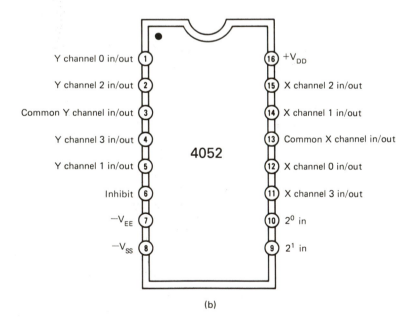

(b)

Fig. 6-19 CMOS multiplexer/demultiplexer pin diagrams: (a) 4051 8-channel, (b) 4052 differential 4-channel pairs, (c) 4053 differential 2-channel pairs.

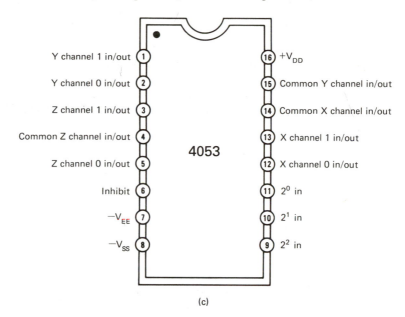

(c)

Fig. 6-19 (*Continued*)

REVIEW

In this chapter we have discussed how to convert from one type of code to another using both discrete logic functions and specific integrated circuits designed for these purposes. Although these functions have been traditionally referred to as either encoders or decoders, we have lumped them together in a single class called code converters. In addition, we have also discussed the multiplexer, whereby we can select a single signal from a number of input signals and pass it on to a single output path. On the other hand, we can go the opposite route by using a demultiplexer to direct a single input signal to one of several possible output paths.

INTRODUCTION TO THE EXPERIMENTS

The following experiments are designed to demonstrate the operation of TTL multiplexers, demultiplexers, and decoders. The experiments that you will perform can be summarized as follows:

Experiment	Purpose
6–1	Demonstrates the operation of the 7442 TTL 4-to-10 line code, or a BCD-to-decimal decoder.
6–2	Demonstrates the operation of a 16-line multiplexer and demultiplexer using the 74150 and 74154 devices.
6–3	Demonstrates the operation of the 7442 device as a 1-of-8 demultiplexer.

EXPERIMENT 6–1

PURPOSE

The purpose of this experiment is to demonstrate the operation of the 7442 TTL 4-to-10 line, or BCD-to-decimal decoder.

**PIN CONFIGURATION OF
INTEGRATED-CIRCUIT DEVICE**

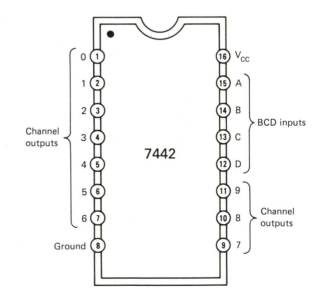

Fig. 6-20 7442 pin diagram.

Required Parts and Equipment

1   7442 TTL 4-to-10 line code converter
4   TTL logic switches
1   LED monitor
1   solderless breadboard
1   TTL power supply

SCHEMATIC DIAGRAM OF CIRCUIT

STEP 1

Wire the circuit shown in the schematic diagram of Figure 6–21. Initially, leave the input to the LED monitor unconnected to any point on the circuit.

STEP 2

Apply power to the breadboard and set all four logic switches at logic 0 so that the 4-bit BCD binary input is *DCBA* = 0000. Now, using the LED monitor sequentially at pins 1, 2, 3, 4, 5, 6, 7, 9, 10, and 11, determine which decoder outputs are logic 0 and 1.

If you have wired the circuit correctly, you should find that all outputs indicate a logic 1, *except pin 1*, which is a logic 0. From the 7442 pin diagram, pin 1 is the channel 0 output. It alone is a logic 0 because the 7442's 4-bit input code equals the decimal number 0.

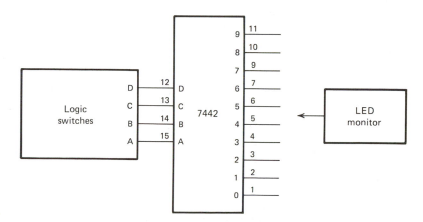

Fig. 6–21 Schematic diagram for Experiment 6–1.

STEP 3

Now set the four logic switches so that the 4-bit BCD binary input is *DCBA* = 0001. Now, again using the LED, monitor sequentially at pins 1, 2, 3, 4, 5, 6, 7, 9, 10, and 11; determine which decoder outputs are logic 0 and 1.

You should now find that all outputs indicate a logic 1, *except pin 2*, which is a logic 0. From the 7442 pin diagram, pin 2 is the channel 1 output. It alone is a logic 0 because the 7442's 4-bit input code equals the decimal number 1.

STEP 4

Now set the four logic switches so that the 4-bit BCD binary input is *DCBA* = 0010. Now, again using the LED, monitor sequentially at pins 1, 2, 3, 4, 5, 6, 7, 9, 10, and 11; determine which decoder outputs are logic 0 and 1.

You should now find that all outputs indicate a logic 1, *except pin 3*, which is a logic 0. From the 7442 pin diagram, pin 2 is the channel 2 output. It alone is a logic 0 because the 7442's 4-bit input code equals the decimal number 2.

You should now conclude that, depending on the 4-bit binary input code, you can select one of ten possible output channels that alone will be 0, while the remaining nine output channels will remain at logic 1. Although any 4-bit code can represent a maximum of 16 possible 4-bit combinations of 1s and 0s, the 7442 decodes only ten of these, that is, those 4-bit binary numbers between 0000 and 1001, which equal the decimal numbers 0 through 9.

STEP 5

Verify that, by setting the four logic switches to any 4-bit binary number code between 0000 and 1001, you are able to decode a single output channel whose output is 0 while the remaining nine channels are 1. You should also find that the decoded output channel is the decimal equivalent of the 4-bit input code to the 7442.

STEP 6

In step 5 you should have concluded that the 7442 decodes only those 4-bit inputs between 0000 and 1001, although a 4-bit number can range from 0000 to 1111, for a total of 16 possible codes. Now set the input switches to the remaining six codes: $DCBA = 1010, 1011, 1100, 1101, 1110,$ and 1111. For each of these 4-bit codes in turn, use the LED monitor to determine which output channels are 0 and 1.

For all six of these 4-bit codes, which represent the decimal numbers 10 through 15, you should have found that *there is no uniquely decoded output channel* whose output is 0, and that *all output channels are 1*. Although the 7442 is a 4-to-10 line decoder, it is for this reason that it is more frequently referred to as a *BCD-to-decimal decoder*. In order to decode all 16 possible channels, you must then use a 74154 device, which is used in the next experiment.

EXPERIMENT 6-2

PURPOSE

The purpose of this experiment is to demonstrate the operation of both a multiplexer and a demultiplexer using the 74150 and 74154 devices.

PIN CONFIGURATION OF
INTEGRATED-CIRCUIT DEVICES

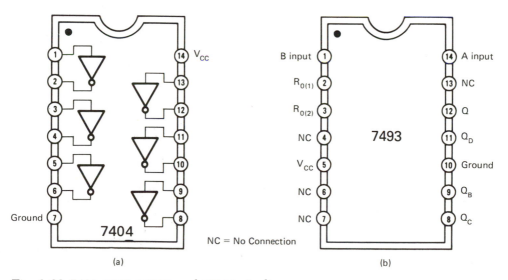

(a)   (b)

Fig. 6-22 7404, 7493, 74150, and 74154 pin diagrams.

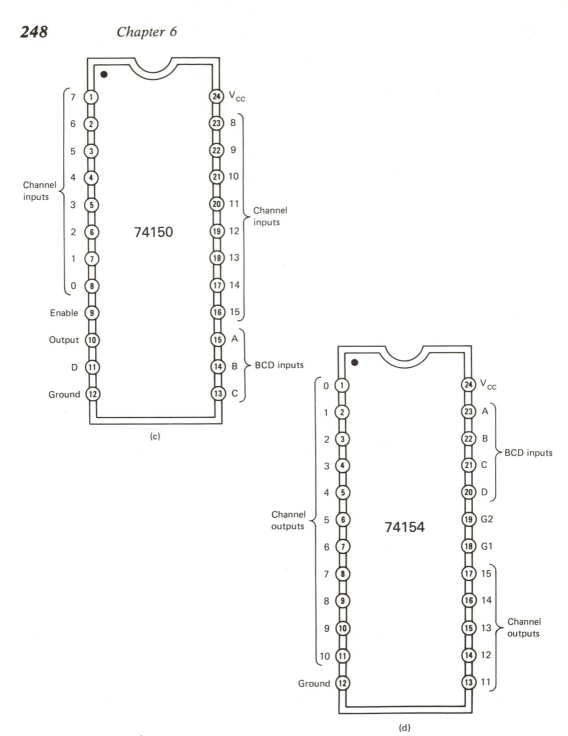

Fig. 6-22 (*Continued*)

Required Parts and Equipment

1   7404 TTL hex inverter
1   7493 TTL 4-bit binary (divide-by-16) counter
1   74150 TTL 16-to-1 line multiplexer
1   74154 TTL 1-to-16 line demultiplexer
1   TTL logic pulser
5   LED monitors
1   square-wave clock (approximately 10 Hz)
1   solderless breadboard
1   TTL power supply

SCHEMATIC DIAGRAM OF CIRCUIT

STEP 1

Wire the circuit shown in the schematic diagram of Figure 6–23. The 74150 is a TTL 16-to-1 line multiplexer, while the 74154 is a 1-to-16 line demultiplexer. The 7493 counter will allow you to cycle through all 16 binary inputs.

STEP 2

Apply power to the breadboard and use the pulser to cycle the 7493 counter so that LED monitors A, B, C, and D all indicate logic 0. Then connect the output of 10 Hz to pin 8 of the 74150 multiplexer and LED monitor E to pin 1 of the 74154 demultiplexer. What do you observe on LED monitor E?

You should observe that LED monitor E flashes repeatedly on and off at the same rate as the square-wave clock signal. Since the 4-bit code at the DATA SELECT inputs of the 74150 equals the decimal number 0, only input channel 0 can receive or *multiplex* this clock signal and pass it on to the single output channel at pin 10 of the 74150. The 74154 then takes this signal and passes it on to the output channel that is determined by the 4-bit code at its DATA SELECT inputs. Since both the 74150 and 74154 have the same 4-bit data select code, the clock signal is transferred or *demultiplexed* to channel 0, as indicated by LED monitor E. You have then transmitted a clock signal on a specific input channel through a multiplexer and received it on a similar channel using a demultiplexer.

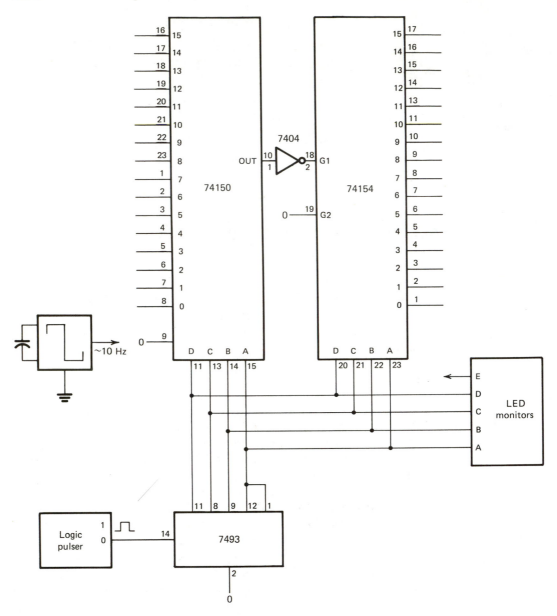

Fig. 6-23 Schematic diagram for Experiment 6–2.

STEP 3

Now connect the output of the square-wave clock to pin 1 of the 74150, which corresponds to input channel 7. What happens to LED monitor E?

When the clock signal is transferred to input channel 7, you should observe that LED monitor E no longer flashes on and off at pin 1 (output channel 0) of the 74154.

STEP 4

Using LED monitor E, check the other 15 output channels of the 74154 demultiplexer to see if the 10-Hz clock signal appears at any other output.

You should find that the other 15 output channels all have outputs that are at logic 1 and that none of the 16 outputs flash on and off at the 10-Hz clock signal.

STEP 5

Now use the pulser to cycle the 7493 counter so that the DATA SELECT inputs for both the 74150 and 74154 are *DCBA* = 0111 on LED monitors A to D. Connect LED monitor E to pin 8 of the 74154. What happens to the LED monitor?

You should find that the LED monitor again flashes on and off at a 10-Hz rate just like the input clock signal. Why?
Since the DATA SELECT input for the 74150 in binary is equal to the decimal number 7 (*DCBA* = 0111), the only signal at the output of the multiplexer (pin 10) is the signal present at input channel 7. Likewise, since the DATA SELECT input for the 74154 in binary is also equal to the decimal number 7, the input signal at pin 18 will only be passed on to the channel 7 output at pin 8.
You have just now selected one input signal from 16 possible input channels and then transferred this to one of 16 other possible signal lines. In general, the 4-bit DATA SELECT inputs to the multiplexer and demultiplexer do not have to be the same. You can transmit data via channel 13 of the multiplexer and output it via channel 4 of the demultiplexer. In this case the DATA SELECT inputs for the 74150 should be *DCBA* = 1101, while the inputs for the 74154 should be *DCBA* = 0100.

EXPERIMENT 6-3

PURPOSE

The purpose of this experiment is to demonstrate the operation of the 7442 device as a 1-of-8 demultiplexer.

PIN CONFIGURATION OF INTEGRATED-CIRCUIT DEVICE

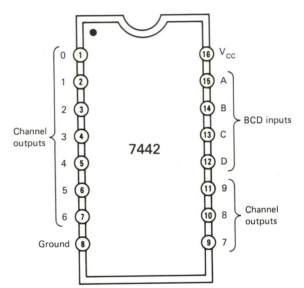

Fig. 6-24 7442 pin diagram.

Required Parts and Equipment

1   7442 TTL 4-to-10 line code converter
3   TTL logic switches
2   LED monitors
1   square-wave clock (approximately 10 Hz)
1   solderless breadboard
1   TTL power supply

SCHEMATIC DIAGRAM OF CIRCUIT

STEP 1

Wire the circuit shown in the schematic diagram of Figure 6–25. The only difference between this circuit and that of Experiment 6–1 is that a clock signal (input data) is now applied to one of the inputs and the other three inputs controlled by logic switches instead of all four inputs controlled by logic switches. Initially, leave the input to LED monitor 2 unconnected to any point on the circuit.

STEP 2

Apply power to the breadboard and set the three logic switches so that *CBA* = 100. What output channel would you expect the 10-Hz train of clock pulses to appear at?

Since the DATA SELECT inputs are *CBA* = 100, equal to the decimal number 4, the train of clock pulses should appear at output channel 4, which is pin 5 of the 7442 decoder.

STEP 3

With the logic switches set at *CBA* = 100, verify that the 10-Hz train of clock pulses appears only at pin 5 (output channel 4). You should also observe that the signal at pin 5 of the 7442 is *not inverted* with respect to the input data at pin 12 (the *D* DATA SELECT input).

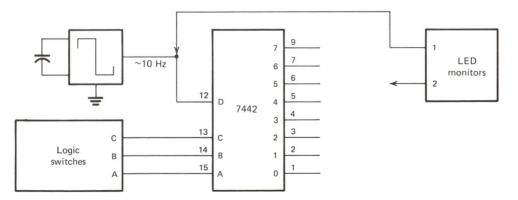

Fig. 6-25 Schematic diagram for Experiment 6–3.

STEP 4

Now set the three logic switches so that $CBA = 001$. What output channel would you expect the 10-Hz train of clock pulses to appear at?

Since the DATA SELECT inputs are $CBA = 001$, equal to the decimal number 1, the train of clock pulses should appear at output channel 1, which is pin 2 of the 7442 decoder.

STEP 5

With the logic switches set at $CBA = 001$, verify that the 10-Hz train of clock pulses appears only at pin 2 (output channel 1).

Although the DATA SELECT inputs are set at $CBA = 001$, why do you also see LED monitor 2 repeatedly flash on and off when placed at output channel 9 (pin 11) of the 7442 decoder?

Since the input train of clock pulses alternates between logic 0 and 1 at the D input, the DATA SELECT inputs then alternate between the 4-bit codes 0001 and 1001, which correspond to the decimal numbers 1 and 9, respectively. When the input clock pulse is at logic 0, then $DCBA = 0001$ so that output 1 (pin 2) is at logic 0 and output 9 (pin 11) is at logic 1. On the other hand, when the input clock pulse is at logic 1, then $DCBA = 1001$ so that output 1 (pin 2) is at logic 1 and output 9 (pin 11) is at logic 0. Consequently, the decoded output channel 1 is *noninverted* with respect to the input data pulse train, while the data at output 9 are *inverted*.

Although we have seen in Experiment 6–1 that the 7442 is manufactured primarily as a BCD-to-decimal decoder, we have shown here that we can also use the 7442 as a 1-of-8 line de-multiplexer by using the D DATA SELECT input to receive the input data pulses. In this case, output channels 8 and 9 are not used as they will produce pulse trains that are inverted with respect to the input data pulse train.

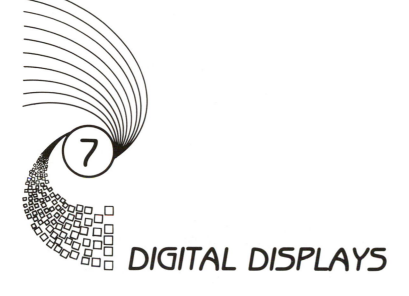

DIGITAL DISPLAYS

INTRODUCTION

For a number of digital devices, the output usually drives some type of visual display, which could be either a single light to indicate the status of a portion of the circuit or a numerical display, such as a digital time clock. Like a picture, digital displays give us visual information about the operation of the circuit. This chapter discusses digital displays as well as how to interface them to various TTL and CMOS devices.

OBJECTIVES

After completing this chapter, you will understand how to:

- *Describe the operation of LED and LCD displays.*

- *Interface single incandescent lamps and LEDs with TTL and CMOS devices.*

- *Interface common-anode and common-cathode seven-segment LED displays with TTL and CMOS decoder/drivers.*

- *Implement leading and trailing zero suppression of multidigit LED displays.*

- *Interface seven-segment LCDs using 74C48 and 4055 devices.*

255

- *Detect display overflow.*

- *Multiplex seven-segment LED displays.*

- *Obtain BCD information from seven-segment displays.*

- *Describe the functions needed to build both a digital time clock and a frequency counter.*

INCANDESCENT LAMPS

The incandescent lamp, often referred to as a light bulb, can require several hundred milliamperes of current even for voltages as low as 1.5 V. Consequently, lamps cannot be directly connected to the outputs of most integrated-circuit devices. However, using a transistor driver circuit, as shown in Figure 7–1, incandescent lamps may be safely used as output indicators.

If a supply voltage V_S is used to power the lamp, the lamp voltage V_L should be

$$V_L = V_S - V_{CE}(\text{sat}) \qquad (7\text{–}1)$$

where $V_{CE}(\text{sat})$ is the collector-to-emitter saturation voltage of the transistor and is typically several tenths of a volt. The circuit functions such that the lamp lights when a logic 1 level is present at the input of the lamp driver. For TTL devices, the minimum output voltage for a logic 1 is typically 2.4 V, while a logic 1 output for CMOS devices is

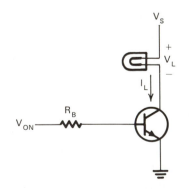

Fig. 7-1 NPN transistor driver for incandescent lamps.

basically equal to the supply voltage. Without derivation, the value of base resistor R_B is found from

$$R_B = \frac{(V_{ON} - 0.7)\beta}{2.5 I_L} \qquad (7\text{--}2)$$

where

$$V_{ON} = \text{minimum logic 1 voltage}$$

$$I_L = \text{lamp current at } V$$

$$\beta = \text{transistor beta (or } H_{fe})$$

As an example, suppose a 5-V, 60-mA lamp is to be connected to a TTL device and powered by the same supply. The transistor used has a dc current gain (beta) of 100. Since V_{ON} for TTL devices is typically 2.4 V, the base resistor is found to be

$$R_B = \frac{(2.4 - 0.7 \text{ V})(100)}{(2.5)(60 \text{ mA})}$$

$$= 1133 \ \Omega$$

LIGHT-EMITTING DIODE

A *light-emitting diode*, or simply *LED*, emits light when it is forward biased. That is, when the anode is more positive than the cathode, current flows through the diode in the direction from anode to cathode. As shown in Figure 7–2, the typical LED has a voltage drop of approximately 1.7 V when forward biased. In addition, the amount of light emitted, which may be colored with either a red, green, yellow, or orange lens, is proportional to the diode's forward current I_F.

Typically, the LED is placed in series with a resistor, which is used to limit the amount of forward current flowing through the LED, which may range from 10 to 20 mA (50% to 100% relative brightness). Assuming 1.7 V is dropped across the LED, the value of the current-limiting resistor is determined from

$$R = \frac{V_S - 1.7}{I_F} \qquad (7\text{--}3)$$

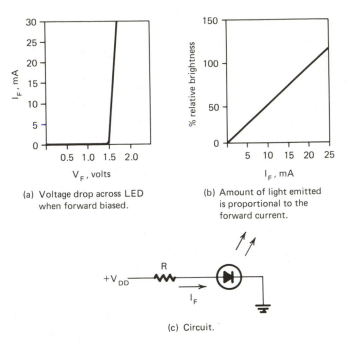

(a) Voltage drop across LED
when forward biased.

(b) Amount of light emitted
is proportional to the
forward current.

(c) Circuit.

Fig. 7-2 LED characteristics: (a) voltage drop across LED when forward biased, (b) amount of light emitted is proportional to the forward current, (c) schematic diagram.

and the power rating of the resistor should be greater than

$$P = I_F^2 R \qquad (7-4)$$

Otherwise, the resistor will overheat and break.

If a LED is connected to the output of a TTL device, which operates from a 5-V supply, the forward LED current must be less than 16 mA, as most TTL devices can only provide a maximum output current of 16 mA without damage to the device. Therefore, the minimum current limiting resistance value is

$$R = \frac{5 - 1.7\ \text{V}}{16\ \text{mA}}$$

$$= 206\ \Omega$$

In practice, a resistance of 330 Ω is typically used.

As shown in Figure 7–3a, a LED is used to indicate when the output of a 2-input TTL NAND gate is a logic 0. When the output of the

NAND gate is logic 0 (or ground), essentially 5 V is dropped across both the LED and the resistor. Since the LED anode is more positive than the cathode, the LED is forward biased and is illuminated. If the output of the NAND gate is a logic 1 (i.e., approximately 5 V), then the anode and cathode voltage are equal and the LED is off. On the other hand, a logic 1 output of the NAND gate can be displayed by a LED by using the scheme of Figure 7–3b. Notice that the polarity of the LED is now reversed.

If a TTL device is to drive a single LED in addition to other TTL devices, the sum of the LED forward current and the total current drawn by the additional devices must be less than 16 mA. Each input of a TTL device typically draws 1.6 mA. If two TTL devices and a single LED are connected to the output of a TTL device, as shown by the 2-input NAND gate of Figure 7–4, then the maximum possible LED forward current is 16 mA − (2 × 1.6 mA), or 12.8 mA.

Because of their limited output current at low supply voltages, most CMOS devices are unable to drive LEDs directly. Figure 7–5 illustrates several practical methods for driving LEDs. Figures 7–5a and b use PNP and NPN transistors, respectively, to drive the LED. The value for the current-limiting resistor is that given by Eq. 7–2. For both circuits, a logic 1 at the transistor's base causes the LED to light. Also, LEDs may be driven from the output of a 4050 CMOS noninverting buffer, as shown in Figure 7–5c.

SEVEN-SEGMENT LED DISPLAYS

Most pocket calculators, frequency counters, digital voltmeters, synthesized CB transceivers, microwave ovens, some AM/FM radios, and

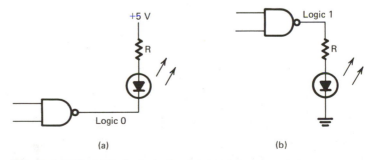

Fig. 7-3 TTL–LED logic indicators: (a) lit when logic 0, (b) lit when logic 1.

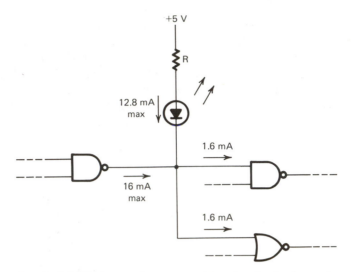

Fig. 7-4 LED forward current with other logic elements connected.

televisions all have something in common. They all use seven-segment LED displays to indicate numbers on their front panel. The name *seven-segment* LED display comes from the fact that there are actually seven *separate* LEDs arranged in a pattern to form any possible number from 0 to 9.

' As shown in Figure 7–6, a seven-segment LED display uses seven LEDs, with each segment location given a specific letter from *a* to *g*. By

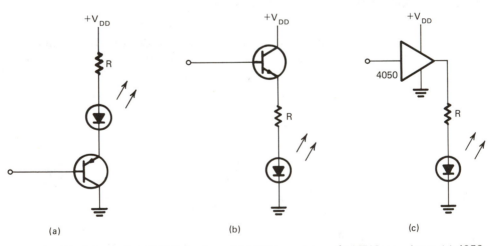

Fig. 7-5 LED drivers for CMOS devices: (a) PNP transistor, (b) NPN transistor, (c) 4050 buffer.

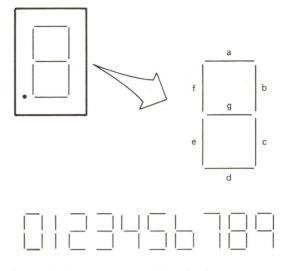

Fig. 7-6 Seven-segment LED display with segment identification.

forward biasing a given combination of LED segments, one can display the numbers 0 through 9. As an example, the number 1 has LED segments *b* and *c* forward biased, while the number 8 has all seven LED segments forward biased. In addition to these seven segments, there is usually an additional LED segment used to serve as a decimal point to the left of the number. Besides the numbers 0 through 9, it is possible to forward bias those LED segments to form other characters. For example, forward biasing segments *a, b, e, f,* and *g* forms the letter P.

Like individual LEDs, seven-segment LED displays come with red, green, yellow, or orange lenses. Depending on the intended application, seven-segment LED display packages come either as a single digit or as a multidigit LED display, which can contain up to as much as ten digits. Whether packaged as a single digit or multidigit display, seven-segment LED displays are categorized as either *common cathode* or *common anode.*

As its name implies, a *common-cathode* seven-segment LED display has the cathodes of all the individual LED segments connected together, as shown in Figure 7–7a. In operation, this common-cathode connection is then grounded while the anode segments are each connected to separate current-limiting resistors or LED drivers, as shown to form the number 1.

On the other hand, a *common-anode* display has the anodes of all the individual LED segments connected together, as shown in Figure

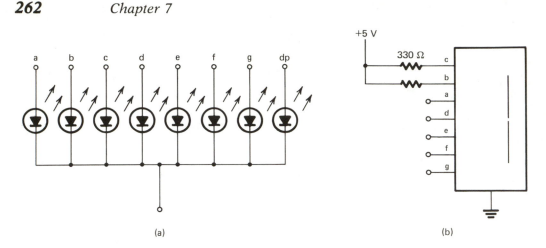

Fig. 7-7 Seven-segment common-cathode LED display: (a) equivalent circuit with decimal point, (b) forward biasing LED segments *b* and *c* to display the number 1.

7–8. In operation, this common anode is connected to the positive logic supply voltage, while the cathode segments are each connected to ground via resistors or LED drivers.

Like single-digit displays, *multidigit* LED displays can be categorized as either common cathode or common anode. Shown in Figure 7–9 is a typical representation for a four-digit common-cathode LED display. Including the decimal point, there should be 4 × 8 or 32 individual LED segments. For each of the four digits, there is a common-cathode connection. However, instead of having 32 separate anode connections, all anodes of the same segment are internally connected in parallel. Consequently, the *a* LED segments of all four

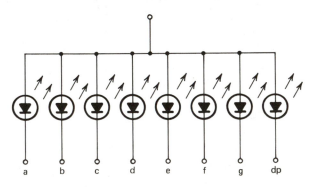

Fig. 7-8 Equivalent circuit for a seven-segment common-anode LED display with decimal point.

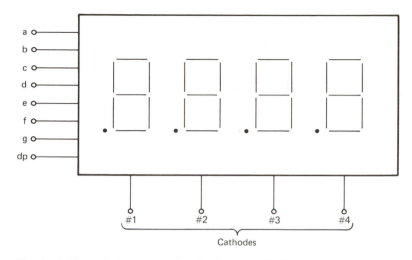

Fig. 7-9 Four-digit common-cathode LED display.

digits are connected together, the *b* segments of all four digits are connected together, and so on. In this manner, we reduce the number of required connections. By grounding the appropriate digit's cathode, we can then select the proper digit. To display numbers on all four digits simultaneously, we have to *multiplex* the display. Display multiplexing is explained later in this chapter.

Besides displaying numbers, LED displays are available that display *alphanumeric* characters. That is, all numbers from 0 to 9, all letters from *a* to *z* or A to Z, and common symbols like !, ", #, $, &, and the like. Figure 7–10a shows a representation of a 5 × 7 alphanumeric display. The display is arranged in a 7-row (cathodes) by 5-column (anodes) matrix plus a decimal point, which is commonly referred to as a *7 × 5 dot matrix*. By forward biasing the proper LED segments, any alphanumeric character can be represented. As an example, Figure 7–10b illustrates how the letter G can be represented on a 7 × 5 dot matrix display.

LIQUID CRYSTAL DISPLAYS

Unlike LED displays, a *liquid crystal display* (LCD) does not emit light. A thin layer of a liquid crystal material is placed between two sheets of glass, which are coated with an electrical conductor that makes them conductive electrodes. The rear glass electrode, sometimes called the *backplane*, is usually transparent, while the front glass electrode is manufactured to form a display, such as a seven-segment figure-8 pattern.

LCDs are manufactured using one of two conventions, transmissive or reflective. In a *transmissive* LCD, the rear glass electrode is made transparent. For the *reflective* LCD, the rear electrode has a mirrorlike film surface. Like LEDs, segment connections are made available. However, LCDs must be used with drivers that are specifically made for them. Since a LCD consumes only a fraction of the power required for a similar LED display, they are intended primarily to be interfaced with CMOS devices.

TTL AND CMOS DISPLAY DRIVERS

Although the lamp and LED display driver circuits discussed earlier in this chapter can be used to interface displays to the outputs of TTL and CMOS circuits, they are frequently more expensive to implement and occupy more space than specific integrated-circuit devices designed for this purpose. Since TTL and CMOS devices generally present different problems, they will be discussed separately in the following sections.

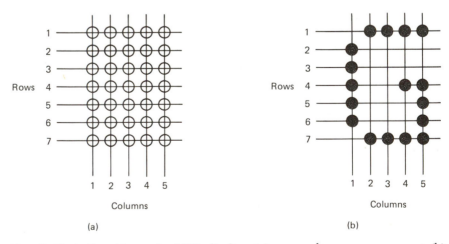

Fig. 7-10 A 5 × 7 matrix LED display: (a) row–column arrangement, (b) displaying the letter G.

TTL Drivers

TTL display drivers generally have a 4-bit BCD input and multiple outputs, whose number depends on what type of display is being driven. As shown in Figure 7–11, the 7445 device is used to drive incandescent lamps having maximum ratings of 80 mA to 30 V, although the 7445 itself must be powered from a 5-V TTL supply. Depending on the BCD inputs, only one of the ten lamps will be lit at any time, as summarized by Table 7–1. The unique decoded output is 0, while the remaining nine outputs are all 1. In effect, the 7445 is a *BCD to 1-of-10 decoder* with each decoded output having its own lamp driver.

For seven-segment LED displays, as shown by the pin diagram in Figure 7–12, the 7447 is used for *common-anode* displays, while the 7448 is used for *common-cathode* types. Internally, both have a BCD-to-7-segment decoder that provides a either a logic 0 (7447) or 1 (7448) output for the segment to be lit. All other outputs are their complement. For each of the seven segments, a current-limiting resistor, typically 330 Ω, is required for the standard 5-V TTL supply, as shown in Figure 7–13. Besides the numbers 0 through 9, both devices produce nonnumerical LED display patterns for equivalent BCD inputs 10

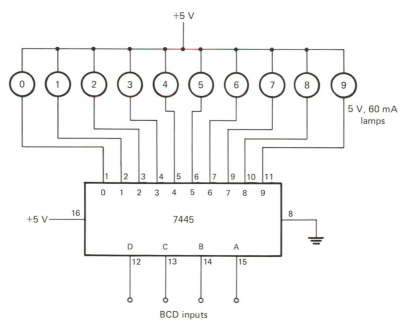

Fig. 7-11 7445 TTL incandescent lamp driver.

Table 7-1
Truth Table for the 7445 Decoder/Driver

BCD Input				Output									
D	C	B	A	0	1	2	3	4	5	6	7	8	9
0	0	0	0	0	1	1	1	1	1	1	1	1	1
0	0	0	1	1	0	1	1	1	1	1	1	1	1
0	0	1	0	1	1	0	1	1	1	1	1	1	1
0	0	1	1	1	1	1	0	1	1	1	1	1	1
0	1	0	0	1	1	1	1	0	1	1	1	1	1
0	1	0	1	1	1	1	1	1	0	1	1	1	1
0	1	1	0	1	1	1	1	1	1	0	1	1	1
0	1	1	1	1	1	1	1	1	1	1	0	1	1
1	0	0	0	1	1	1	1	1	1	1	1	0	1
1	0	0	1	1	1	1	1	1	1	1	1	1	0

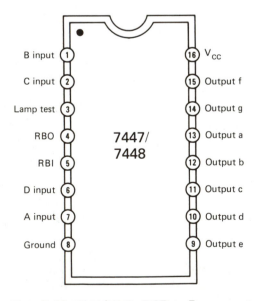

Fig. 7-12 7447/7448 BCD-to-7 segment
decoder/driver pin diagram.

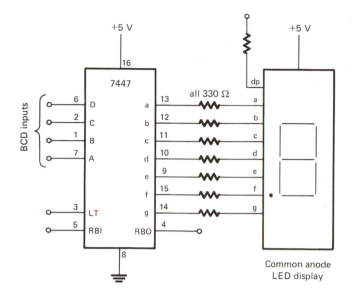

Fig. 7-13 7447–common anode display interface.

through 14, while the display is blanked (no segments lit) for the equivalent BCD input 15. The truth tables for the 7447 and 7448 decoder/drivers are given by Tables 7–2 and 7–3, respectively.

Besides the BCD inputs and seven-segment outputs, both devices have *zero suppression* logic when used with two or more displays. Zero suppression of digital displays is used to blank out unwanted zeros. For example, if a seven-digit display without zero suppression reads 00345.80, then by implementing the zero suppression feature the two zeros before the 3 and the zero after the 8 are blanked out, and the display would read 345.8. Blanking of the zeros at the beginning of the significant numbers is called *leading zero suppression*, while blanking of the end zeros is called *trailing zero suppression*.

The RBI, or *ripple-blanking input*, and RBO, or *ripple-blanking output*, signals are used to blank the display. Figure 7–14 shows how either the 7447 or 7448 is connected for leading zero suppression for a five-digit display. Note that the RBI pin of the most significant digit is tied at logic 0, and the RBO pin is connected to the RBI of the next lower order digit, and so on. Figure 7–15 illustrates the required connections for trailing zero suppression.

The final feature of both the 7447 and 7448 is the LT or *lamp test* input. By placing a logic 0 to this input, all seven LED segments are lit, displaying the number 8. For the 7447, all outputs are 0, while outputs

Table 7-2
Truth Table for the 7447 Decoder/Driver for Common-Anode LED Displays

Displayed Number	BCD Input				Seven-Segment Output							Segments Lit
	D	C	B	A	a	b	c	d	e	f	g	
0	0	0	0	0	0	0	0	0	0	0	1	a, b, c, d, e, f
1	0	0	0	1	1	0	0	1	1	1	1	b, c
2	0	0	1	0	0	0	1	0	0	1	0	a, b, d, e, g
3	0	0	1	1	0	0	0	0	1	1	0	a, b, c, d, g
4	0	1	0	0	1	0	0	1	1	0	0	b, c, f, g
5	0	1	0	1	0	1	0	0	1	0	0	a, c, d, f, g
6	0	1	1	0	1	1	0	0	0	0	0	c, d, e, f, g
7	0	1	1	1	0	0	0	1	1	1	1	a, b, c
8	1	0	0	0	0	0	0	0	0	0	0	a, b, c, d, e, f, g
9	1	0	0	1	0	0	0	1	1	0	0	a, b, c, f, g
—	1	0	1	0	1	1	1	0	0	1	0	d, e, g
—	1	0	1	1	1	1	0	0	1	1	0	c, d, g
—	1	1	0	0	1	0	1	1	1	0	0	b, f, g
—	1	1	0	1	0	1	1	0	1	0	0	a, d, f, g
—	1	1	1	0	1	1	1	0	0	0	0	d, e, f, g
Blank	1	1	1	1	1	1	1	1	1	1	1	None

are 1 for the 7448. The LT provides a simple feature to determine if there are any burned-out LED segments.

With or without zero suppression, the following situation arises. Suppose that a five-digit display reads 16932. Without leading zero suppression, we have no way of knowing if the actual number is larger than 16932 or not. The five-digit display can only display a number that is between 00000 and 99999. If the actual number is 516932, the display would not indicate the most significant number, 5. In this case, a *display overflow* exists. By using a flip-flop following the last decade counter, as shown in Figure 7–16, the output pulse from the last counter triggers it when the number exceeds the display's capability.

Table 7–3
Truth Table for the 7448 Decoder/Driver for Common-Cathode LED Displays

Displayed Number	BCD Input				Seven-Segment Output							Segments Lit
	D	C	B	A	a	b	c	d	e	f	g	
0	0	0	0	0	1	1	1	1	1	1	0	a, b, c, d, e, f
1	0	0	0	1	0	1	1	0	0	0	0	b, c
2	0	0	1	0	1	1	0	1	1	0	1	a, b, d, e, g
3	0	0	1	1	1	1	1	1	0	0	1	a, b, c, d, g
4	0	1	0	0	0	1	1	0	0	1	1	b, c, f, g
5	0	1	0	1	1	0	1	1	0	1	1	a, c, d, f, g
6	0	1	1	0	0	0	1	1	1	1	1	c, d, e, f, g
7	0	1	1	1	1	1	1	0	0	0	0	a, b, c
8	1	0	0	0	1	1	1	1	1	1	1	a, b, c, d, e, f, g
9	1	0	0	1	1	1	1	0	0	1	1	a, b, c, f, g
—	1	0	1	0	0	0	0	1	1	0	1	d, e, g
—	1	0	1	1	0	0	1	1	0	0	1	c, d, g
—	1	1	0	0	0	1	0	0	0	1	1	b, f, g
—	1	1	0	1	1	0	0	1	0	1	1	a, d, f, g
—	1	1	1	0	0	0	0	1	1	1	1	d, e, f, g
Blank	1	1	1	1	0	0	0	0	0	0	0	None

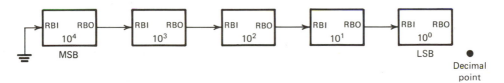

Fig. 7–14 7447 decoder/driver leading display zero suppression.

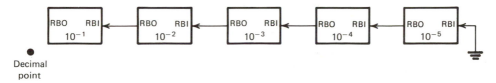

Fig. 7–15 7447 decoder/driver trailing display zero suppression.

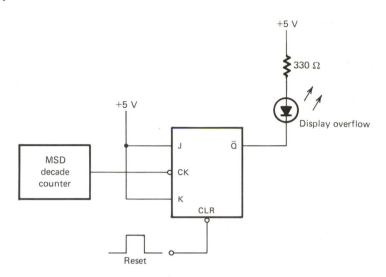

Fig. 7-16 Display overflow indicator.

CMOS Drivers

Perhaps the easiest method to enable most CMOS devices to drive a wide variety of seven-segment LED displays is to use an integrated-circuit transistor array, such as the CA3081 or CA3082, as shown in Figure 7–17. The CA3081 consists of a series of seven high-current transistors in a common-emitter configuration for use with *common-anode* displays, while the CA3082 is a common-collector array used with *common-cathode* displays.

For supply voltages greater than 3.5 V, Figure 7–18 shows how the CA3082 array is used with a common-cathode display and a 4026 or 4033 CMOS decade counter that has decoded seven-segment outputs. Each anode is connected to a current-limiting emitter resistor, whose value is determined from

$$R_E = \frac{V_{DD} - 2.4}{I_F} \qquad (7\text{--}5)$$

The 2.4-V term in the numerator of Eq. 7–5 is collectively the typical base–emitter voltage drop of the transistor (0.7 V) plus the forward voltage drop of the LED segment (1.7 V). On the other hand, Figure 7–19 shows a similar arrangement, but using the CA3081 array with a

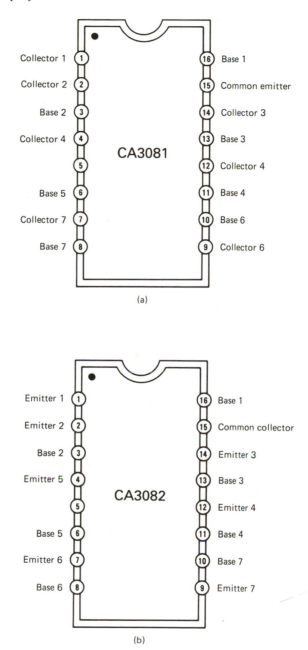

Fig. 7-17 Transistor driver array pin diagrams for 7-segment LED displays: (a) CA3081 common-emitter array, (b) CA3082 common-collector array.

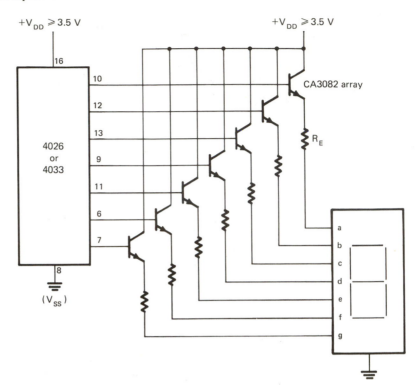

Fig. 7-18 Common-collector transistor array, common-anode LED display interface for common-cathode displays.

common-anode display. The collector resistor R_C for each segment is determined from

$$R_C = \frac{V_{DD} - 1.9}{I_F} \qquad (7\text{--}6)$$

where the 1.9-V term in the numerator takes into account the forward voltage drop of the LED segment (1.7 V) and the collector-to-emitter saturation voltage of the transistor (0.2 V). The maximum value for each base resistor R_B is

$$R_B \max = \frac{(V_{DD} - 0.7)\beta_{\min}}{I_F} \qquad (7\text{-}7)$$

For both the arrays, the minimum value of β is typically 30.

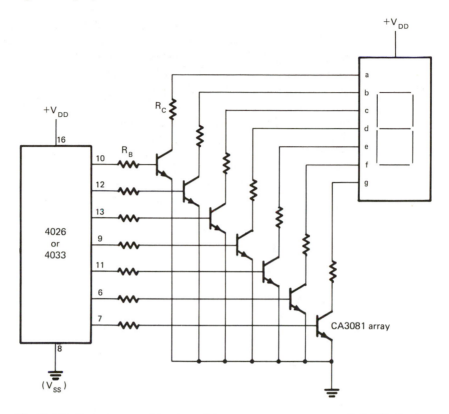

Fig. 7-19 Common-emitter transistor array, common-anode LED display interface.

For both the 4026 and 4033 counters, the decoded output for the numbers 6 and 9 differs from that produced by the 7447 and 7448 TTL decoder/drivers. As illustrated in Figure 7–20, the decoded output produces a "tail" on the numbers 6 and 9. For the number 6, the *a* segment is added, while the *d* segment is added for the number 9.

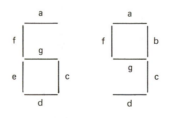

Fig. 7-20 Decoded numbers 6 and 9 with "tail."

The major difference between the 4026 and 4033 counters is that the 4026 has a *display enable* input, which is used to turn off the display when at logic 0. The 4033, on the other hand, does not have this display enable feature, but like the 7447/7448, has nonsignificant zero suppression using the RBI and RBO pins as shown in Figure 7–21.

Another CMOS decoder/driver is the 4511 shown in the pin diagram of Figure 7–22. Besides having an internal BCD-to-7-segment decoder for common-cathode displays, it also has a 4-bit latch to retain the 4-bit BCD code once this information is removed from the inputs. Its outputs, unlike the 4026 and 4033, require no transistor arrays and are capable of driving the LED display directly through current-limiting resistors, as shown in Figure 7–23. The resistor value is found from

$$R = \frac{V_{DD} - 1.7}{I_F} \qquad (7\text{–}8)$$

If the CMOS supply voltage is 9 V and the LED forward current is 15 mA, the series resistor is

$$R = \frac{V_{DD} - 1.7}{I_F}$$

$$= \frac{9 - 1.7 \text{ V}}{15 \text{ mA}}$$

$$= 487 \text{ } \Omega$$

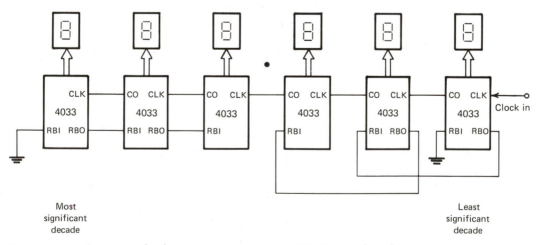

Fig. 7-21 Leading zero display suppression using 4033 CMOS decade counters.

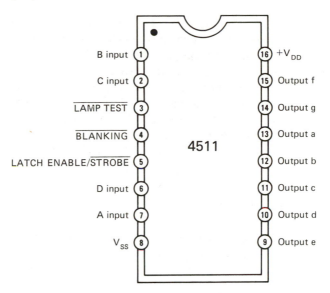

Fig. 7-22 4511 CMOS BCD-to-7-segment display decoder/driver pin diagram.

for which a 470-Ω resistor can be used. The power dissipation of the resistor should also be determined. By Ohm's law,

$$P = I_F^2 R \qquad\qquad (7\text{--}9)$$
$$= (15 \text{ mA})^2 (470 \ \Omega)$$
$$= 0.11 \text{ watt (W)}$$

so that a 1/4-W resistor can be used safely.

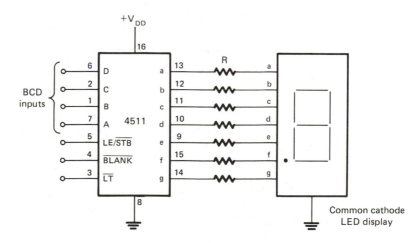

Fig. 7-23 4511–common cathode display interface.

Although the 4511 is primarily intended for common-cathode displays, a common-anode display can be used with the CA3081 transistor array, as shown in Figure 7–24, using Eqs. 7–6 and 7–7 to determine the resistor values.

For liquid crystal displays, the 74C48 can also be used for LCDs, as shown in Figure 7–25. Unlike LED displays, external excitation is required for the LCD in the form of a square-wave signal with a frequency between 30 to 200 Hz applied to the LCD's backplane. Besides the 74C48, the 4055 also can be used, as shown in Figure 7–26. It contains a BCD-to-7-segment decoder and driver.

Multiplexed Displays

For LED displays requiring four or more digits, it is usually more advantageous to multiplex or strobe the display. In general, display

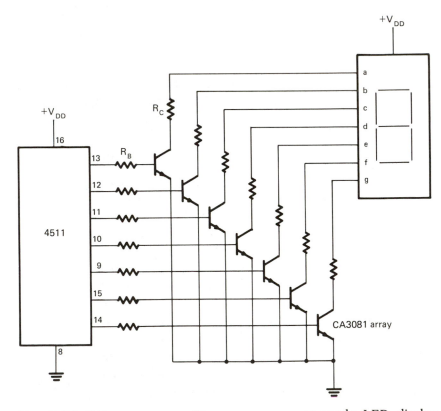

Fig. 7-24 4511, common-emitter array, common-anode LED display interface.

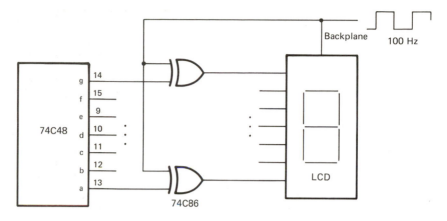

Fig. 7-25 CMOS–LCD decoder/driver using a 74C48.

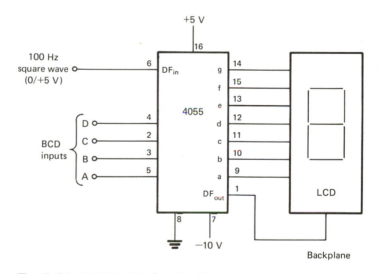

Fig. 7-26 CMOS–LCD decoder/driver using a 4055.

multiplexing is a rather efficient method that uses a single decoder/ driver for all the digits, as illustrated by the diagram of Figure 7–27. The single decoder/driver is sequentially time shared with each of the display digits, one digit at a time. Since all the like segments are wired in parallel, each digit out of N total digits will be enabled for $1/N$ the time it takes to strobe all N digits.

If the display is multiplexed without any increase in the forward current to each of the seven LED segments, the resultant display will be

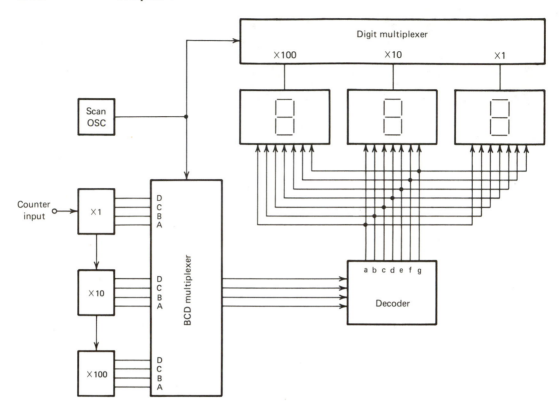

Fig. 7-27 Display multiplexing block diagram.

dim. This is because the *average* current will now be less than that without multiplexing. The average forward LED current is

$$I_F \text{ (average)} = \frac{I_F}{N} \qquad (7\text{--}10)$$

By increasing the current to N times the normal value, the average current will then be sufficient to provide a visible display.

The *scan rate* or strobing frequency must be high enough so that there is no noticeable flicker as each digit is sequentially enabled. The human eye can distinguish light pulses for frequencies up to about 16 Hz. Above this, light pulses tend to appear as a continuous glow, but vary somewhat among individuals and viewing and ambient lighting conditions.

OBTAINING BCD INFORMATION FROM SEVEN-SEGMENT DISPLAYS

For a number of applications, particularly microcomputer applications, it may be appropriate to obtain the equivalent 4-bit code corresponding to the number shown on each digit of a series of seven-segment LED displays. Few CMOS-type digital clocks, panel meters, pocket calculators, and counters, such as the 4026 and 4033, have provisions for obtaining BCD data.

Using a 74C915 CMOS decoder, as shown in Figure 7–28, we are able to convert the seven-segment information (Table 7–4) into BCD code. The states of the seven-segment inputs can be selected by the INVERT/NONINVERT control input. A logic 0 at this input selects active HIGH true decoding at the seven-segment inputs. The ERROR output is 1 whenever a nonstandard seven-segment code is detected. When the LATCH ENABLE (LE) input is 1, the BCD output is latched via *tri-state* outputs.

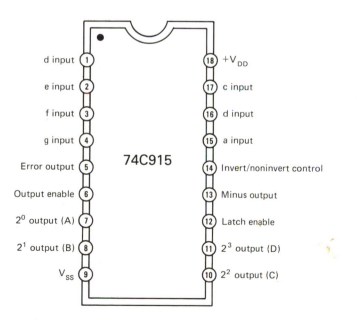

Fig. 7-28 74C915 CMOS 7-segment-to-BCD decoder.

Table 7-4
Truth Table for the 74C915 Seven-Segment-to-BCD Decoder

Displayed Number	Seven-Segment Input							BCD Output				ERROR Output	MINUS Output
	a	*b*	*c*	*d*	*e*	*f*	*g*	*A*	*B*	*C*	*D*		
0	1	1	1	1	1	1	0	0	0	0	0	0	0
1	0	1	1	0	0	0	0	0	0	0	1	0	0
2	1	1	0	1	1	0	1	0	0	1	0	0	0
3	1	1	1	1	0	0	1	0	0	1	1	0	0
4	0	1	1	0	0	1	1	0	1	0	0	0	0
5	1	0	1	1	0	1	1	0	1	0	1	0	0
6	0	0	1	1	1	1	1	0	1	1	0	0	0
6[a]	1	0	1	1	1	1	1	0	1	1	0	0	0
7	1	1	1	0	0	0	0	0	1	1	1	0	0
8	1	1	1	1	1	1	1	1	0	0	0	0	0
9	1	1	1	0	0	1	1	1	0	0	1	0	0
9[a]	1	1	1	1	0	1	1	1	0	0	1	0	0
Blank	0	0	0	0	0	0	0	1	1	1	1	0	0
Minus (−)	0	0	0	0	0	0	1	×	×	×	×	1	1

[a]6 or 9 with "tail."

× represents tri-state condition.

PUTTING IT ALL TOGETHER

At this point, we have discussed the basic elements needed to build most digital devices. Due to advances in technology, many functions that would ordinarily require a large number of gates, flip-flops, counters, and so on, are now provided by some manufacturers as a single integrated-circuit device, some of which are discussed in Chapter 10. Nevertheless, to appreciate how these various logic functions can be connected together, this section focuses on the design of a simple digital time clock and a frequency counter using TTL devices with LED displays.

Digital Time Clock

A simple digital time clock can be built using 7490 and 7492 counters as shown in Figure 7–29. Counters $U1$ and $U2$ divide the 60-Hz power line frequency by 60 down to 1 pulse each second. Counters $U3$ and $U4$ count the output pulses from $U2$, which is the elapsed time in seconds from 00 to 59, and repeat this every 60 clock cycles, or every 60 seconds (s). Counters $U5$ and $U6$ function the same as counters $U3$ and $U4$, but this time they count the time in minutes, receiving 1 pulse each minute from $U4$. Since there are 60 s in every minute, when counters $U4$ and $U3$ both reset to zero after every 60 input clock pulses, counter $U5$ increments its count by 1, and so on, until the minutes count reaches 59.

The output from $U6$ has a rate of 1 pulse per hour. Unlike counters $U3$ through $U6$, the counter function that would display the hours *never displays 00*. The displayed numbers from the hours is any number from 01, 02, through 12, but *never* 00.

J–K flip-flops $U7$ through $U10$ are connected as a BCD synchronous BCD counter. After the BCD count reaches 1001 (i.e., a 9), the next input pulse recycles the counter to zero. At the same time, flip-flop $U10$ toggles the last flip-flop of the hours counter ($U11$) so that it displays the number 1, for a combined hours count of 10. The next two pulses reaching $U7$ advance the displayed count from 10 to 12. The very next input pulse to $U7$ attempts to advance flip-flops $U7$ to $U10$ to display a 3 (0011) count. Since the Q output of flip-flops $U7$, $U8$, and $U11$ are all logic 1, the output of the NAND gate is 0, which immediately resets flip-flops $U8$ to $U11$. Only flip-flop $U7$ remains untouched, whose Q output is still 1, so that the hours display is 01 and the cycle starts over again.

From this basic circuit, we can add additional features to enhance the utility of the time clock, for example, an alarm to go off at a predetermined time. Figure 7–30 shows one of many approaches that may be taken. Although not completely shown, a series of six 7485 4-bit comparators is used to compare the predetermined time for the alarm with the actual BCD codes that are instantaneously being displayed. When the alarm time (in BCD) equals the displayed time, the $A = B$ output of the last 7485 comparator is 0 for 1 s. This negative pulse triggers a 555 timer wired as a 5-min monostable multivibrator. The output of the monostable enables another 555 timer, which now functions as an audio oscillator, whose output drives a speaker. Once sounded, the alarm may be silenced by momentarily grounding the reset pin on the monostable timer.

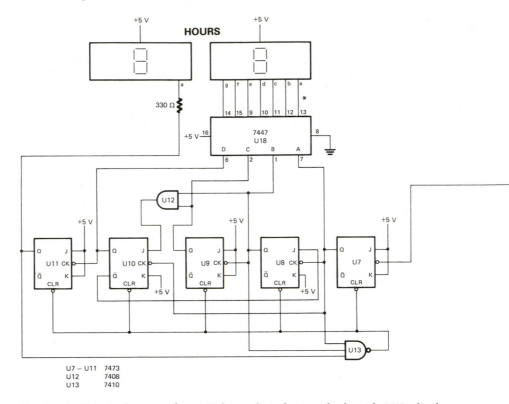

Fig. 7-29 Circuit diagram for a 12-hour digital time clock with LED display.

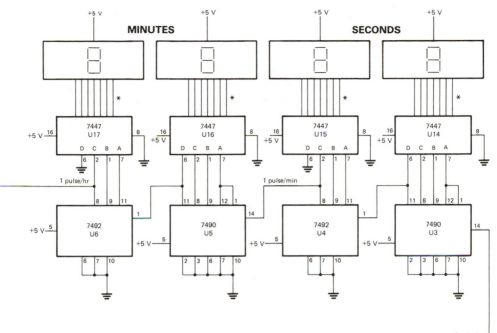

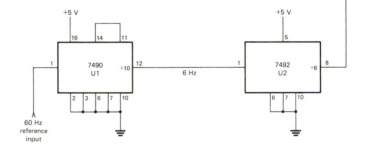

*330 Ω resistors for all
segments omitted for clarity.

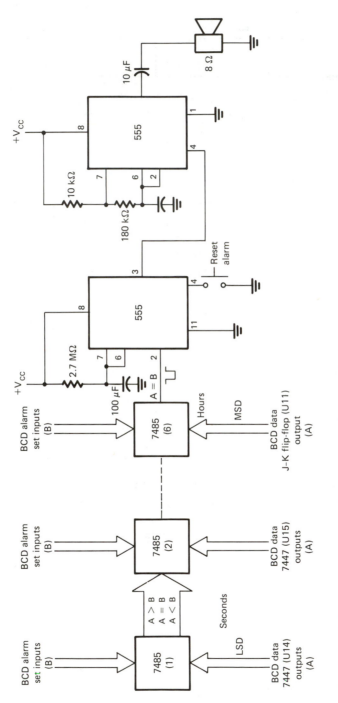

Fig. 7-30 Scheme for addition of an alarm to the time clock.

As a practical matter, it is not now economically feasible to build a digital time clock from scratch using gates, flip-flops, and counters such as we have done here. This is because there are already many inexpensive devices that have all these functions plus an audible alarm and other features, in a single integrated-circuit device that occupies a much smaller space than the 18 logic devices used in this example.

2. Digital Frequency Counter

Figure 7–31 illustrates a block diagram as an aid in understanding what makes a frequency counter tick. Like the digital time clock, the heart of all frequency counters is the *frequency reference*, which is frequently called a *time base*. This is a very stable, fixed-frequency clock signal that controls the internal operations of the counter circuitry. This stable signal is then separated into three other signals that tell the circuitry to (1) first reset all decade counters to zero, (2) then count the number of pulses of the unknown frequency for a fixed amount of time, and (3) then store (latch) this total count to update the digital display. After the circuits complete these three sequences, this cycle begins all over again. In actuality, a frequency counter is nothing more

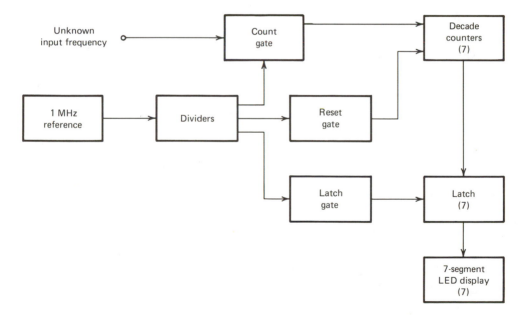

Fig. 7-31 Frequency counter block diagram.

than an *events per unit time* device. We merely count the number of input pulses for a given amount of time. Since this time window is typically 1 s long, the number of input pulses for each 1 s of time is the frequency of that signal and has the unit of hertz (Hz).

• Frequency Reference

The frequency reference can be as simple as using the 60-Hz power line frequency or can be made more accurate by using a crystal-controlled oscillator, such as those discussed in Chapter 4. As shown in Figure 7–32, a 1-MHz TTL crystal oscillator is formed using 7400 NAND gates *U1* and *U2*. NAND gate *U3* buffers the oscillator's output from the loading effects of the stages to follow.

The remainder of the frequency reference consists of five cascaded 7940 divide-by-10 counters to divide the 1-MHz oscillator

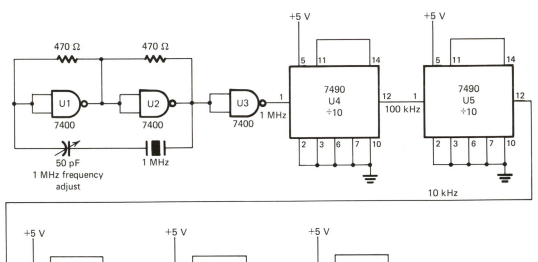

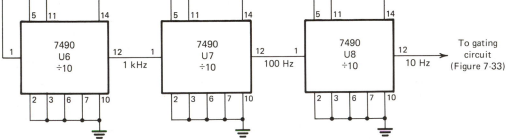

Fig. 7-32 10-Hz frequency counter crystal-controlled time base reference.

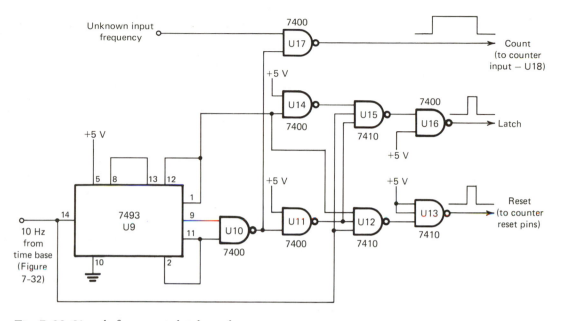

Fig. 7-33 Signals for count, latch, and reset.

frequency down to 10 Hz, which provides the reference for the count, reset, and latch timing signals.

• Count, Reset, and Latch Timing

A circuit for providing the proper sequence of count, reset, and timing signals from the 10-Hz reference is shown in Figure 7–33. The 7493 (*U*9) divides the 10-Hz reference by 12 (in BCD) to control the count gate (*U*17), the reset gate (*U*13), and the latch gate (*U*16).

As illustrated by its associated timing diagram of Figure 7–34, the *reset* gate (*U*13) output first is at logic 1 for 0.05 s; then the *count* gate is enabled for 1 s, allowing the unknown input frequency to pass through. Finally, the *latch* gate (*U*16) is disabled to freeze the display until the next cycle is completed.

• Count, Latch, and Display Decades

The counter, latch, and display circuit of Figure 7–35 is a seven-decade display capable of displaying frequencies up to 9,999,999 Hz. The arrangement is a straightforward approach without display multiplexing. Although not completely shown, there are seven identical 7490

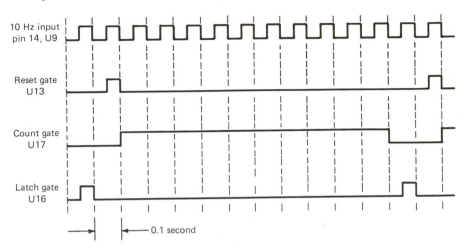

Fig. 7-34 Timing diagram.

decade (BCD) counters, 7475 4-bit latch, 7447 BCD-to-7-segment decoder/driver, and common-anode LED display sequences, one for each of the seven decades. The required inputs for this circuit are the count, reset, and latch outputs from Figure 7–33. In addition, the 7447 decoder/drivers have been arranged to provide leading zero suppression.

In all, 45 separate logic gates, counters, latches, decoders, and displays are required for this circuit, although the actual number of TTL device packages is slightly less, since the 7400 and 7410 gates are manufactured in multigate packages. A similar design arrangement can be easily implemented using CMOS devices. However, as described in Chapter 10, one can create an equivalent, but physically smaller frequency counter using only *two* CMOS devices plus a seven-digit LED display.

REVIEW

As most devices are not capable of adequately driving digital displays directly, this chapter discussed how to properly interface the outputs of TTL and CMOS devices to a variety of digital displays. The display can be as simple as a single LED to indicate a logic 0 or 1 state, or the display can indicate alphanumeric information, such as the output of a digital counter.

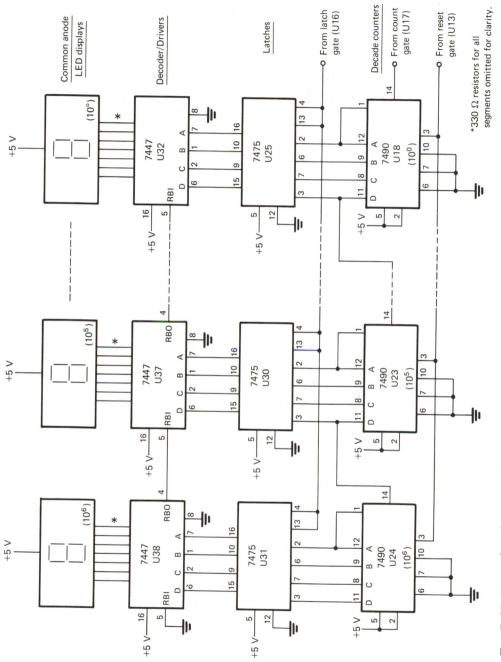

Fig. 7-35 Seven-decade counter, latch, and display.

A series of seven-segment digital displays can be driven individually or, to save costs, multiplexed with the aid of decoders. Furthermore, this chapter discussed how to enhance the appearance of digital displays by removing or blanking out nonsignificant zeros. As examples of how digital displays are used with the various logic elements discussed in previous chapters, this chapter presented possible circuits for a 12-hour digital time clock and a 60-MHz frequency counter.

INTRODUCTION TO THE EXPERIMENTS

The following three experiments are designed to familiarize you with the interface and operation of the LED and the seven-segment LED display. The experiments that you will perform can be summarized as follows:

Experiment	Purpose
7–1	Demonstrates the operation of a single LED used to indicate the output state of a TTL device.
7–2	Demonstrates how to light specific segments of a common-anode seven-segment LED display.
7–3	Demonstrates the use of a seven-segment LED display to visually indicate the output of a decade counter.

EXPERIMENT 7–1

PURPOSE

The purpose of this experiment is to demonstrate the operation of a single LED to indicate the output state of a TTL device, such as a 7476 flip-flop.

PIN CONFIGURATION OF
INTEGRATED-CIRCUIT DEVICE AND LED

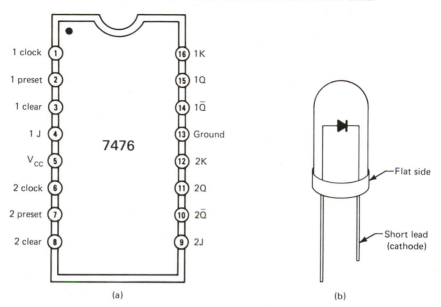

(a)   (b)

Fig. 7-36 7476 and LED pin diagrams.

Required Parts and Equipment

1   330–Ω resistor
1   LED (red, green, orange, or yellow lens)
1   7476 TTL dual *J–K* flip-flop
2   TTL digital pulsers
1   solderless breadboard
1   dc voltmeter
1   TLL power supply

SCHEMATIC DIAGRAM OF CIRCUIT

STEP 1

Wire the circuit shown in the schematic diagram of Figure 7–37. Make sure that you have wired the LED correctly. The LED's *cathode* terminal is the one closest to the *flattened* side of the LED case. If you are using a LED whose leads you know have not been cut or shortened,

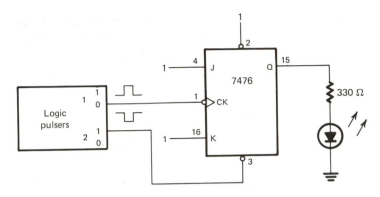

Fig. 7-37 Schematic diagram for Experiment 7–1.

the cathode lead is the *shorter* of the two. You can easily destroy the LED if you accidently have the two leads reversed.

STEP 2

After you are sure that everything is wired correctly, apply power to the breadboard. Press and release pulser 2 so that the 7476 flip-flop will be *reset* (i.e., the Q output at pin 15 is a logic 0). With your dc voltmeter, measure the voltage at pin 15 of the 7476 with respect to ground. It should be approximately 0 V, which corresponds to a logic 0 state. Is the LED lit or unlit?

> Since you have just reset the flip-flop, the Q output at pin 15 is 0. Both the anode and cathode LED terminals are essentially at the same voltage (i.e., ground or 0 V). Consequently, no current flows through the LED from its anode-to-cathode terminals, so the LED is *unlit.*

STEP 3

Now press and release pulser 1 to toggle the flip-flop. From Chapter 3, you should now know what the Q output of the flip-flop should be. What do you now observe about the LED?

> You should observe that the LED is lit.

Why?

> By triggering the flip-flop once, output Q (pin 15) now goes from logic 0 to 1. As a general rule, let's assume that the voltage that

corresponds to a logic 0 is equal to the positive supply voltage, or +5V. Since its anode is now more positive than its cathode, the LED is then said to be *forward biased,* so current flows from its anode to the cathode.

STEP 4

Now place your dc voltmeter's *positive* lead at the LED's anode and the meter's negative lead at the LED's *cathode* terminal. Measure the voltage drop across the LED and record your result:

$$V_d = \underline{\hspace{2cm}} \text{volts}$$

Using Ohm's law, you should now be able to determine how much current is flowing through the LED when it is forward biased (Eq. 7–3). As an example, we measured 1.6 V, which means that approximately 5 − 1.6 or 3.4 V is dropped across the 330-Ω series resistor. Consequently, there is 3.4 V/330 Ω or a current of 10.3 mA flowing through the LED when lit.

The 330-Ω resistor therefore limits the LED current to 10.3 mA. By increasing the LED current, we can increase the brightness of the LED display. However, too much current will destroy the LED the same as if we had connected the LED backward.

As is discussed in the next chapter, the maximum output current that can be obtained from most standard TTL devices is 16 mA. Consequently, the maximum LED current when connected to the output of a standard TTL device is also 16 mA, which, for our experiment, says that the minimum series resistance required is 3.4 V/16 mA, or 213 Ω.

STEP 5

Now press and release pulser 1 several times. The LED should flash on and off as the flip-flop is toggled. When the Q output is at logic 1, the LED is lit.

STEP 6

Now disconnect the power from the breadboard and wire the LED as shown in Figure 7–38. Again, be sure that you have wired the LED correctly.

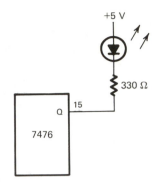

Fig. 7-38 Schematic diagram for Experiment 7–1, step 6.

STEP 7

Apply power to the breadboard and then press and release pulser 1 to reset the 7476 flip-flop. What now happens to the LED?

If you have done everything correctly, the LED should now be lit, even though the *Q* output of the flip-flop is now 0. To be sure that the *Q* output is at logic 0, measure the output voltage at pin 15 with respect to ground. It should be approximately zero, indicating a logic 0 state.

In this step, we have wired the LED so that the anode is at +5 V. When the flip-flop's *Q* output is at logic 0, there will be approximately 5 V across the resistor and the LED. Since the anode is at a higher potential than the cathode, the LED is forward biased and therefore is lit, which, in this case, indicates a *logic 0* state.

STEP 8

Again press and release pulser 1. The LED should now be *unlit*, which indicates a logic 1.

You have just demonstrated that, depending on how the LED is connected, a lit LED can be made to indicate either a logic 0 or 1 state. The general practice, however, is to have a lit LED indicate a *logic 1* state.

EXPERIMENT 7-2

PURPOSE

The purpose of this experiment is to demonstrate how to light specific segments of a common-anode seven-segment LED display.

PIN CONFIGURATION OF SEVEN-SEGMENT LED DISPLAY

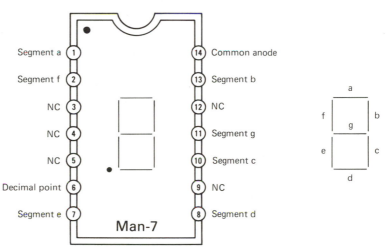

Fig. 7-39 MAN-7 LED display pin diagram.

Required Parts and Equipment

- 7   220-Ω resistors
- 1   MAN-7 common-anode seven-segment LED display (or equivalent)
- 1   solderless breadboard
- 1   TTL power supply

SCHEMATIC DIAGRAM OF CIRCUIT

STEP 1

Wire the circuit shown in the schematic diagram of Figure 7–40. The MAN-7 seven-segment LED display is a *common-anode* type, in that the anodes of all the LED segments are internally tied together and

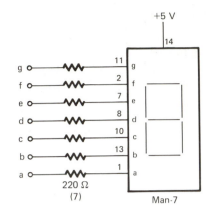

Fig. 7-40 Schematic diagram for Experiment 7-2.

available as a common connection at pin 14. The cathodes of the individual LED segments are separate, requiring a series resistor for each LED segment.

STEP 2

Apply power to the breadboard. Since none of the series resistors are connected to either ground (logic 0) or +5 V (logic 1), the display should be blank. Now ground the resistors that are connected to the *a, b, d, e,* and *g* inputs of LED display. What do you see?

If you have done everything correctly, you should see the number 2 displayed.

STEP 3

For the combinations shown in Table 7–5, light the indicated LED segments and record the resultant number that is displayed.

If you performed every step correctly, you should have displayed the numbers 0, 1, 2, 3, 4, 5, 6, 7, 8, and 9 in that order. You should now be able to display any number from 0 to 9 by forward biasing the proper LED segments.

STEP 4

In addition to the numbers 0 through 9, it is also possible to display most capital and lowercase letters by forward biasing the proper LED segments. For example, a "b" is formed by forward biasing the *c, d, e, f,* and *g* LED segments, while a "P" is formed by the segments *a, b, e, f,* and *g.* As an exercise, try to form the following letters on the seven-segment LED display:

Capitals: A, C, E, F, H, I, J, L, O, P, S, and U
Lowercase: b, c, d, g, h, i, l, o, r, and u

For some of these letters you will have to be somewhat creative.

EXPERIMENT 7-3

PURPOSE

Demonstrates the use of a common-anode seven-segment LED display with a 7447 decoder/driver to visually indicate the output of a 7490 decade counter.

Table 7-5

LED Segments	Displayed Number
a, b, c, d, e, f	
b, c	
a, b, d, e, g	2
a, b, c, d, g	
b, c, f, g	
a, c, d, f, g	
c, d, e, f, g	
a, b, c	
a, b, c, d, e, f, g	
a, b, c, f, g	

PIN CONFIGURATION OF INTEGRATED-CIRCUIT DEVICES
AND SEVEN-SEGMENT LED DISPLAY

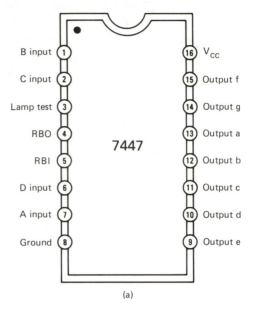

(a)

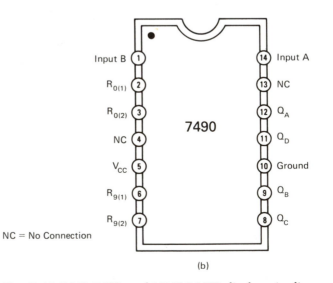

NC = No Connection

(b)

Fig. 7-41 7447, 7490, and MAN-7 LED display pin diagrams.

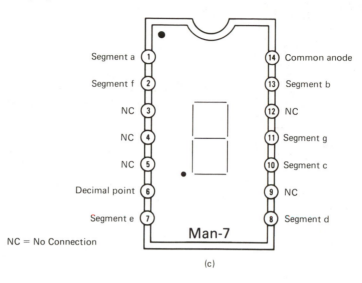

Segment a — 1
Segment f — 2
NC — 3
NC — 4
NC — 5
Decimal point — 6
Segment e — 7

14 — Common anode
13 — Segment b
12 — NC
11 — Segment g
10 — Segment c
9 — NC
8 — Segment d

Man-7

NC = No Connection

(c)

Fig. 7-41 (*Continued*)

Required Parts and Equipment

7 220-Ω resistors
1 7447 TTL BCD-to-7-segment decoder/driver
1 7490 TTL decade counter
1 MAN-7 common-anode seven-segment
   LED display (or equivalent)
4 LED monitors
2 TTL digital pulsers
1 solderless breadboard
1 TTL power supply

SCHEMATIC DIAGRAM OF CIRCUIT

STEP 1

Wire the circuit shown in the schematic diagram of Figure 7–42. Be careful that none of the resistor leads touch each other.

STEP 2

Apply power to the breadboard. Initially, the display might show any number from 0 to 9 as the internal flip-flops of the 7490 decade

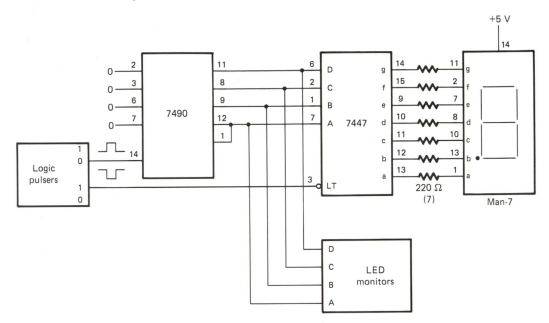

Fig. 7-42 Schematic diagram for Experiment 7-3.

counter might assume any one of ten possible 4-bit output combinations. Is there a relationship between the number on the seven-segment LED display and the status of the four LED monitors?

The decimal equivalent of the 4-bit code displayed by the LED monitors should equal the number shown on the seven-segment display.

STEP 3

Press and release pulser 1 several times. What happens to the LED monitors and the seven-segment display?

You should see that the LEDs count upward. The LED monitors count upward in BCD code, while the seven-segment LED display displays numbers that increase by 1 each time the pulser is released. By continuing to activate pulser 1, the seven-segment LED display counts toward 9 and then starts over again: 0, 1, 2, 3, and so on.

The 7447 decoder/driver takes the 4-bit BCD inputs and decodes them into the proper LED segments for that 4-bit number. As an example, if the BCD inputs were *DCBA* = 0100, which equals the number 4, then the 7447 decodes this so that only the *b, c, f,* and *g* outputs will be at logic 0. Consequently, these four LED segments of the common-anode display are forward biased and therefore will be lit to display the number 4.

STEP 4

Activate pulser 1 until the number 7 is displayed. Now press and release pulser 2. Cycle the counter to display the number 1 and again press in pulser 2. What happens to the seven-segment LED display each time pulser 2 is pressed in (pin 3 of the 7447 at logic 0)?

When pin 3 of the 7447 decoder is at logic 0, *all seven segments of the LED display are lit, and it is independent of the status of the BCD inputs.* This input is called the *lamp test* input and is primarily used to check whether any of the seven LED segments are burned out.

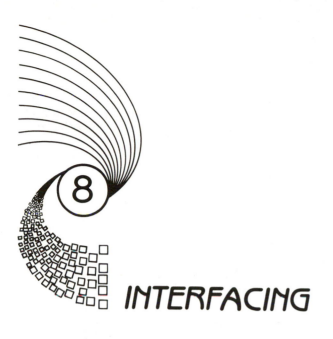

INTERFACING

INTRODUCTION

Both TTL and CMOS devices are capable of being connected to or interfaced with other circuits or logic families. For this to happen, we require that (1) the logic levels be compatible, (2) both logic families operate from a single supply voltage, (3) the devices be kept at safe power dissipation levels, and (4) the noise immunity of the logic families be maintained.

OBJECTIVES

After completing this chapter, you will be able to do the following:

- *Describe suitable methods for the interfacing of standard TTL and CMOS devices.*

- *Describe the input and output interfacing requirements of TTL and CMOS devices.*

303

DIFFERENT LOGIC FAMILIES

When interfacing TTL and CMOS devices, careful attention must be paid to the specific voltage levels that correspond to logic 0 and logic 1.

Standard TTL Input/Output Characteristics

As illustrated in Figure 8–1, the maximum allowable input voltage, V_{IL} max, to ensure a logic 0 input level for most standard TTL devices (the 74 series) is 0.8 V. To ensure a logic 1 input level, the minimum input voltage, V_{IH} min, is 2.0 V. On the output side, the maximum allowable output voltage, V_{OL} max, for a logic 0 is 0.4 V, although this is typically 0.22 V. The minimum output voltage, V_{OH} min, that corresponds to a logic 1 level is 2.4 V.

Each standard 7400 series TTL device is capable of delivering a current of 16 mA to a load, which may be another TTL device,

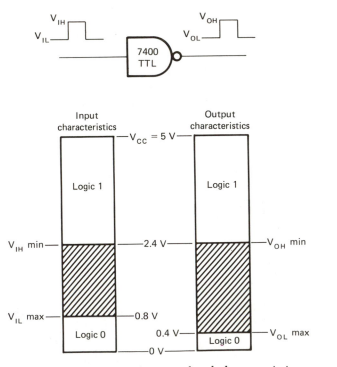

Fig. 8-1 TTL input and output level characteristics.

transistor, or whatever. Typically, a standard TTL device will draw a current of 1.6 mA by each of its inputs. Consequently, it is then possible to connect up to ten standard TTL devices (i.e., 16mA/1.6 mA = 10) to a given output without damage or loss of reliability. This maximum number of devices that may be simultaneously connected to the output of another device of the same logic family is termed the *fan-out* capability, unless noted otherwise for different logic families.

Low-Power Schottky TTL Input/Output Characteristics

In Figure 8–2, the maximum allowable input voltage to ensure a logic 0 input level for most low-power Schottky TTL devices (the 74LS series) is 0.8 V. To ensure a logic 1 input level, the minimum input voltage is 2.4 V. It should be noted that both input levels are identical to standard TTL. However, the maximum allowable output voltage for a logic 0 is

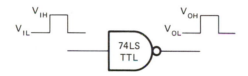

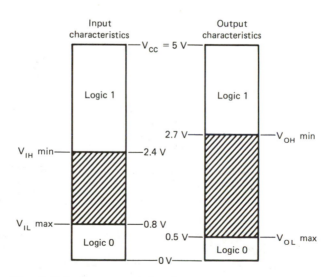

Fig. 8-2 Low-power Schottky TTL input and output level characteristics.

0.5 V. Furthermore, the minimum output voltage that corresponds to a logic 1 level is up to 2.7 V.

Each 74LS series TTL device is capable of delivering a current of 8 mA to a load. Typically, a 74LS TTL device will draw a current of 0.36 mA by each of its inputs. Consequently, it is possible to connect up to 22 74LS TTL devices to a given 74LS device output without damage, which implies a 74LS fan-out capability of 22. If standard TTL devices are connected to a 74LS device, the fan-out becomes 8 mA/1.6 mA, or 5.

CMOS Input/Output Characteristics

Unlike TTL devices, data sheets for CMOS devices do not directly specify the input voltage requirements, but instead give a parameter called *noise immunity*. As illustrated in Figure 8–3, any input voltage to a CMOS device less than 30% of the supply voltage V_{DD} is considered at

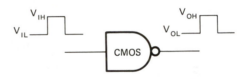

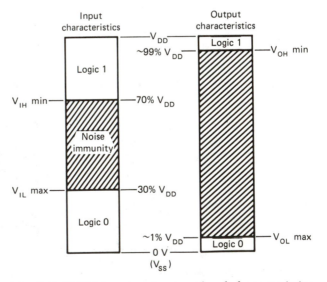

Fig. 8-3 CMOS input and output level characteristics.

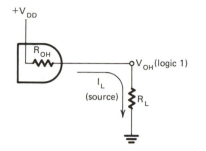

Fig. 8-4 CMOS current source equivalent circuit.

logic 0; any input voltage greater than 70% of V_{DD} is at logic 1. Consequently, a given input voltage can rise to 30% of V_{DD} or fall to 70% of V_{DD} without a change of logic level at the output of the device. The *noise immunity* is then from 30% to 70% of V_{DD}. For a 5-V supply, this corresponds to levels of 1.5 to 3.5 V, or a difference of 2.0 V. Naturally, the noise immunity will increase with increasing supply voltage. At 15 V, for example, the noise immunity is the difference of 10.5 minus 4.5, or 6.0 V.

The maximum output *sink current* and *source current* ratings vary somewhat with different CMOS devices, and, in addition, vary from manufacturer to manufacturer for the same device. As shown in the equivalent circuit of Figure 8–4, the CMOS device acts as a *current source* when the output is at logic 1. The internal output resistance R_{OH} then causes the output to be less than the supply voltage according to the following voltage-divider relationship:

$$V_{OH} = V_{DD} \frac{R_L}{R_L + R_{OH}} \tag{8–1}$$

For a logic 1 output, the output resistance is typically in the 200- to 800-Ω range. When one CMOS device drives another CMOS device, the input impedance is typically 10^{11} Ω (100 GΩ). In this case, the output voltage of the current source is essentially that of the supply voltage. For a typical CMOS-to-CMOS connection, the source current is

$$I_L \text{ (source)} = \frac{V_{OH}}{R_L}$$

$$= \frac{V_{DD}}{R_L + R_{OH}} \tag{8–2}$$

which is, on the average, 10 pA. The maximum source current is typically in the 0.14- to 0.25-mA range for $V_{DD} = 5$ V; 0.35 to 0.6 mA for $V_{DD} = 10$ V; and 1.0 to 1.8 mA for $V_{DD} = 15$ V.

As a current sink, using the equivalent circuit of Figure 8–5, the output voltage is then at logic 0. The internal output resistance, R_{OL}, which is essentially that of R_{OH}, causes the output voltage to be equal to

$$V_{OL} = V_{DD} \frac{R_{OL}}{R_L + R_{OL}} \qquad (8\text{–}3)$$

while the output sink current is

$$I_L \text{ (sink)} = \frac{V_{OL}}{R_{OL}}$$

$$= \frac{V_{DD}}{R_L + R_{OL}} \qquad (8\text{–}4)$$

which is, on the average, 10 pA. The maximum sink current is typically in the 0.4- to 0.8-mA range for $V_{DD} = 5$ V; 0.9 to 1.8 mA for $V_{DD} = 10$ V; and 2.0 to 3.6 mA for $V_{DD} = 15$ V. Although for CMOS devices the fan-out capability is virtually infinite, it is limited to 50 as a general design practice.

INTERFACING CMOS TO TTL

The interfacing of CMOS to TTL devices requires several compromises. Although both logic families will work at the required 5-V TTL supply voltage, the maximum operating speed of the CMOS device is

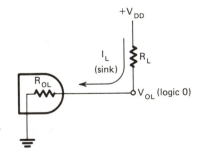

Fig. 8-5 CMOS current sink equivalent circuit.

minimized, being typically 2 MHz, compared to 10 MHz and up for TTL. Virtually all CMOS manufacturers imply that their devices are compatible with TTL. However, this does not mean that the CMOS device can be directly connected to or substituted for a single TTL device (e.g., a 74C74 substituted for a 7474), but rather that both CMOS and TTL will operate at the same 5-V supply voltage. In actuality, *the maximum sink current of most CMOS devices is not sufficient to drive even one standard TTL load.* Remember, a CMOS device can only provide a maximum of 0.8 mA at a supply voltage of 5 V, whereas a standard TTL input draws 1.6 mA.

Another consideration is *logic level incompatibility,* as illustrated in Figure 8–6. Even if the CMOS device were able to provide the necessary 1.6 mA required by a standard TTL device for a logic 0, Eq. 8–4 with R_{OL} equal to 800 Ω tells us that the output voltage of the CMOS device would be

$$V_{OL} = I_L R_{OL} \qquad\qquad (8\text{--}5)$$
$$= (1.6 \text{ mA}) (800 \ \Omega)$$
$$= 1.28 \text{ V}$$

which exceeds the maximum allowable input voltage (V_{IL}) required for a logic 0 for standard TTL devices. For low-power Schottky devices, the input voltage from Eq. 8–5 gives 0.29 V, which is sufficient for a logic 0.

To overcome the limited output sink current of the CMOS device, one of the two arrangements shown in Figure 8–7 should be used when connecting a CMOS device to a standard TTL load. The 4049 (CMOS) is an inverting buffer/converter, while the 4050 is the noninverting counterpart. In either case, all CMOS circuitry that is connected to the input of the buffer can be powered with a 3- to 15-V supply, even though the buffer and the TTL load are operated from a 5-V supply. Both the 4049 and 4050 are designed with input protection circuitry, thus allowing their input voltage to exceed their supply voltage. This buffer arrangement is capable of driving a maximum of two standard TTL loads.

INTERFACING TTL TO CMOS

When interfacing TTL to CMOS loads (Figure 8–8), there are two possibilities:

- The CMOS supply voltage is the same as the TTL supply (i.e., 5V).
- The CMOS supply voltage is between 5 and 15 V.

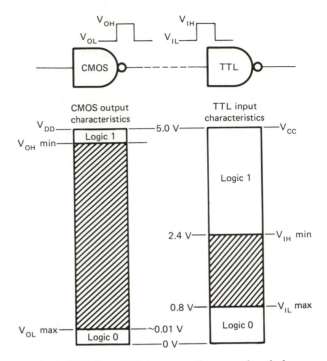

Fig. 8-6 CMOS-to-TTL input and output level characteristics.

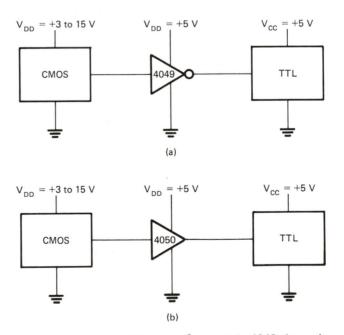

Fig. 8-7 CMOS-to-TTL interfaces: (a) 4049 inverting buffer, (b) 4050 noninverting buffer.

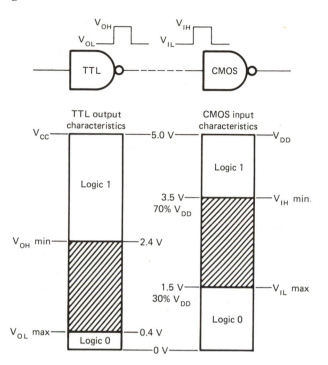

Fig. 8–8 TTL-to-CMOS input and output level characteristics.

In either case, the guaranteed minimum TTL output logic 1 voltage of 2.4 V (2.7 V for low-power Schottky TTL) is less than the minimum CMOS input voltage required for an acceptable logic 1 input, which is 3.5 V.

To overcome the incompatible logic levels, an external pull-up resistor R, as shown in Figure 8–9, is used. To compute the value for this pull-up resistor, the following factors must be known or assumed:

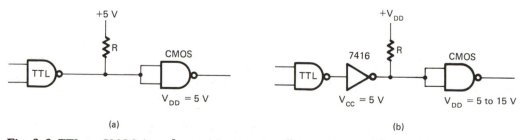

Fig. 8–9 TTL-to-CMOS interfaces: (a) using a pull-up resistor with equal TTL and CMOS supply voltages, (b) for CMOS supply voltages greater than 5 volts.

- Maximum logic 0 collector voltage and current of TTL device, V_{OL} max and I_{OL} max.
- Maximum logic 1 collector-to-emitter leakage current of TTL device, I_{CEO} max.
- The supply voltages, V_{CC} (TTL) and V_{DD} (CMOS).

From these factors, the allowable minimum/maximum pull-up resistance range is determined from

$$R_{min} = \frac{V_{DD} - V_{OL} \text{ max}}{I_{OL} \text{ max}} \qquad (8\text{--}6)$$

and

$$R_{max} = \frac{V_{CC} - V_{IH} \text{ min}}{I_{CEO} \text{ max}} \qquad (8\text{--}7)$$

As an example, let's consider the interfacing of a single TTL device with a single CMOS device, as was shown in Figure 8–9. Typically, the maximum allowable TTL logic 0 collector current is 1.6 mA, and the TTL logic 1 collector-to-emitter leakage current is 100 μA. Therefore,

$$R_{min} = \frac{5 \text{ V} - 0.4 \text{ V}}{16 \text{ mA}}$$
$$= 288 \ \Omega$$

and

$$R_{max} = \frac{5 \text{ V} - 3.5 \text{ V}}{100 \ \mu A}$$
$$= 15 \text{ k}\Omega$$

When interfacing a low-power Schottky device to drive a CMOS load, the pull-up resistance is typically in the 820-Ω to 10-kΩ range. An alternative approach for both standard and low-power Schottky TTLs is to use an NPN transistor as illustrated in Figure 8–10.

BUS AND OPEN-COLLECTOR LOGIC DEVICES

Frequently, it is either desired or required to connect two or more logic functions or whole systems together so that they share common

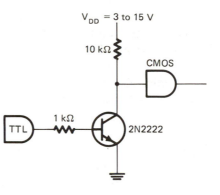

Fig. 8-10 TTL-to-CMOS interface using an NPN transistor.

electrical paths. Such may be the case, as shown by the block diagram of Figure 8–11a, when devices *A* and *B* provide two output signals via a common pair of wires or paths to device *C*, which in turn processes the signals from *A* and *B*. This common two-wire path is termed a *bus*, and its purpose is to minimize the number of interconnections and circuitry required to transfer signals.

In logic block diagrams, the bus is generally represented by a wide line with arrowheads, which indicate the direction of the signals (Figure 8–11b); as such, it is often used in diagrams of computer systems. In general, logic rules dictate that the outputs from *A* and *B* cannot be simultaneously present on the bus without possible damage to some or all of the circuitry. Consequently, the outputs from *A* and *B* must be *isolated* from each other. In our arsenal of integrated-circuit

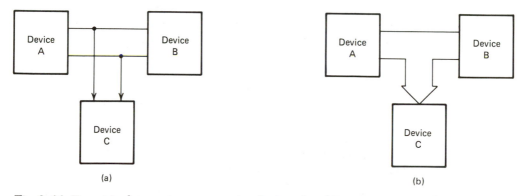

Fig. 8-11 Two-wire bus system connecting devices *A* and *B* to *C*: (a) using individual wires, (b) schematic representation.

devices, we basically have three broad categories of devices that will permit the simultaneous connection of the outputs of two or more devices to form a bus. These are

- Open-collector logic devices
- Three-state logic devices
- CMOS transmission gates

All three types will be discussed in this chapter, starting with open-collector devices in this section.

As briefly discussed in Chapter 1, TTL gates, such as the 7400 NAND gate, have two output transistors connected in *totem-pole* fashion. The benefits of this arrangement are low output impedance and faster switching speed.

Instead of a totem-pole output scheme, a number of TTL family devices have an *open-collector* output, as shown by the equivalent circuit of one section of a 7401 2-input open-collector quad NAND gate in Figure 8–12a. Transistor Q4, which would have been the "lower" transistor of the totem-pole pair, now has its collector open and will not function properly unless an external pull-up resistor is added between the collector lead and the 5-V supply voltage (Figure 8–12b). Table 8–1 summarizes the most commonly available TTL open-collector devices.

As a consequence, open-collector devices are inherently slower than their active totem-pole counterparts, which is their main disadvantage. However, their primary usefulness lies in the fact that outputs of open collectors can be wired together and connected by a common pull-up resistor, which is typically 2.2 kΩ.

As an example, Figure 8–13 shows how all six sections of a 7405 hex open-collector inverter are connected to a common pull-up resistor. If one or more of the inverter inputs are at logic 1, then the common output is logic 0. In other words, this arrangement looks like a conventional *6-input NOR* gate; only when *all* inputs are logic 0 is the output at logic 1.

On the other hand, Figure 8–14a shows two sections of a 7401 quad 2-input open-collector NAND gate connected to the pull-up resistor. In this case, each NAND gate is first treated as if it were an *AND* gate, and the outputs of these two gates are then *ORed together*. Finally, the result is then *inverted*, as summarized by the truth table of Figure 8–14b. In actuality, this arrangement has the same truth table as a type 7451 2-input, 2-wide, AND-OR-invert gate, as was previously discussed in Chapter 2 (Figure 2–32a).

If the output of either NAND gate is 0, then $Q = 0$. In the circuits of Figures 8–13 and 8–14, the output of each logic element is *ORed* with

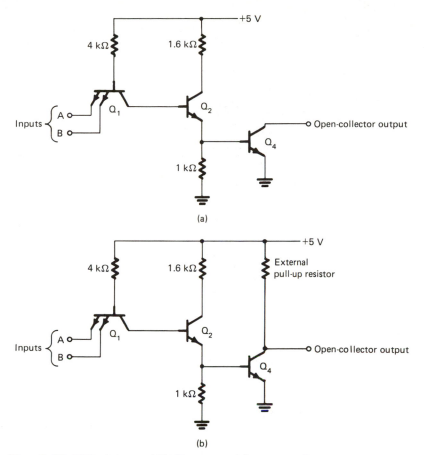

(a)

(b)

Fig. 8-12 TTL 2-input NAND gate with open-collector output: (a) equivalent circuit, (b) equivalent circuit with external pull-up resistor.

Table 8-1
TTL Open-Collector Devices

TTL Device	Function	Open-Collector Equivalent of:
7401	Quad 2-input NAND gate	7400
7405	Hex inverter	7404
7409	Quad 2-input AND gate	7408
7412	Triple 3-input NAND gate	7410
7415	Triple 3-input AND gate	7411
7422	Dual 4-input NAND gate	7420
7433	Quad 2-input NOR gate	7402
74136	Quad exclusive-OR gate	7486/74135
74266	Quad exclusive-NOR gate	74135

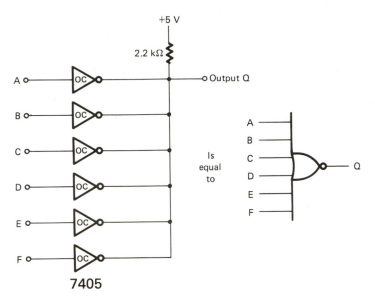

Fig. 8-13 Six inverters with open-collector outputs form a 6-input NOR gate.

the remaining functions and the resultant output inverted. Because of this ORing action, both circuits are examples of why this arrangement is often referred to as either a *wired-OR* or *implied-OR* connection.

As you can probably guess, open-collector devices are not too popular with logic designers as, in addition to being slow, they are difficult to troubleshoot. The use of three-state devices, described in the following section, overcomes problems associated with open-collector devices.

THREE-STATE DEVICES

To overcome the undesirability of open-collector devices, *three-state* logic devices are commonly used. The term "three-state" is the generic term given to the National Semiconductor trademarked name, *Tri-State.®* The uniqueness of three-state logic is reaffirmed by its name: unlike conventional logic, there are *three possible output states.* They are

- A logic 0 state
- A logic 1 state
- A state in which the output looks like an open circuit

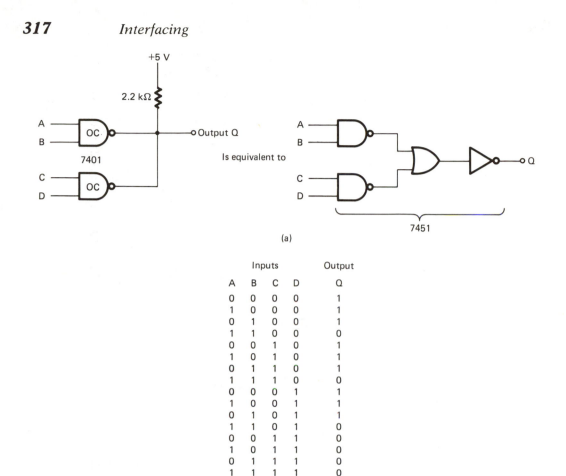

(a)

Inputs				Output
A	B	C	D	Q
0	0	0	0	1
1	0	0	0	1
0	1	0	0	1
1	1	0	0	0
0	0	1	0	1
1	0	1	0	1
0	1	1	0	1
1	1	1	0	0
0	0	0	1	1
1	0	0	1	1
0	1	0	1	1
1	1	0	1	0
0	0	1	1	0
1	0	1	1	0
0	1	1	1	0
1	1	1	1	0

(b)

Fig. 8-14 Two NAND open-collector gates in a wired-OR connection to form a 2-input, 2-wide, AND–OR–invert gate: (a) circuit, (b) truth table.

All three-state TTL devices have an *output enable* or *control* input, which permits the device either to function normally (enabled) or the output signal to be disconnected (disabled) from the rest of the circuit. Depending on the particular device, a logic 0 or 1 may enable the device.

As examples, two of the most frequently used three-state devices are the 74125 and 74126 quad buffers (formerly DM8093 and DM8094), whose pin diagrams are shown in Figure 8–15. Both are noninverting buffers, but the 74125's output is enabled with a logic 0 (note the inversion circle), while the 74126 is enabled with a logic 1, as

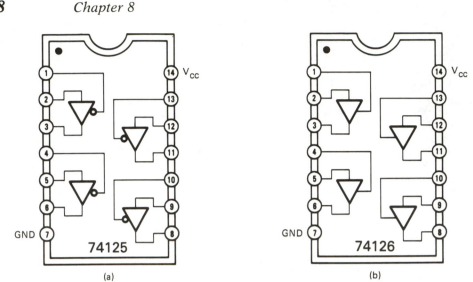

Fig. 8-15 TTL three-state buffer pin diagrams: (a) 74125, (b) 74126.

summarized by the truth table of Table 8–2. In addition, *each buffer section is enabled or disabled independently from the remaining sections.*

As an example, a typical 1-wire bus is shown in Figure 8–16 using four sections of a 74126 buffer. Note that the buffer outputs are connected together onto the bus. The truth table of Table 8–3 summarizes this arrangement. If none of the buffers is enabled (all enable inputs at logic 0), the bus is not connected to anything except inputs to other devices or circuits connected to this bus. Whenever a logic 1 is applied to the enable input of a *single* buffer, the *selected* buffer passes its data on to the bus. The condition in which more than one buffer is enabled at the same time is not allowed, since those buffers are simultaneously trying to pass data on to the bus, and the enabled buffers do not have their outputs isolated from each other.

Table 8-2
74125/74126 Three-State Buffer Truth Table

Device	Output Enable	Output Disable
74125	0	1
74126	1	0

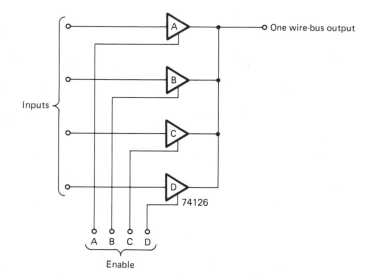

Fig. 8-16 Four-channel multiplexer (bus) using 74126 three-state buffers.

Other frequently used TTL three-state buffers are the 74365 (DM8095), 74366 (DM8096), 74367 (DM8097), and the 74368 (DM8068). Their respective pin diagrams are shown in Figure 8–17.

TRANSMISSION GATES

The CMOS transmission gate or bilateral switch, initially discussed in Chapter 2, is used widely in the switching of both analog and digital

Table 8-3
Truth Table for a Four-Device Three-State Bus

Enable Input				
A	*B*	*C*	*D*	Bus Content
0	0	0	0	Undetermined
1	0	0	0	Data *A*
0	1	0	0	Data *B*
0	0	1	0	Data *C*
0	0	0	1	Data *D*
1	1	1	1	Not allowed

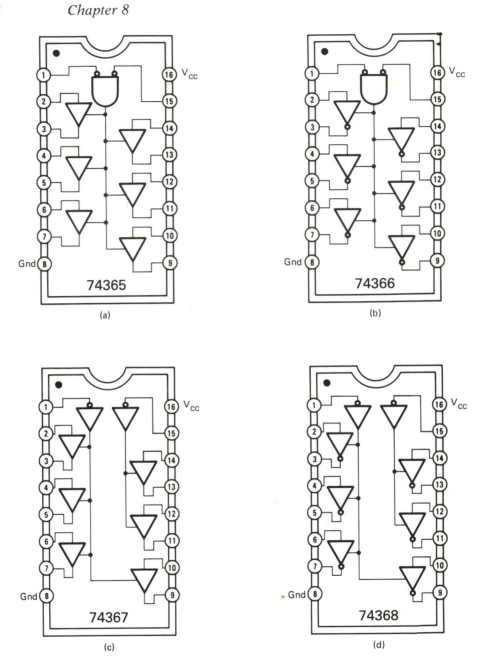

Fig. 8-17 TTL three-state buffer pin diagrams: (a) 74365, (b) 74366, (c) 74367, (d) 74368.

signals to other circuits at rates approaching 10 MHz; it also is a basic building block in the formation of CMOS flip-flops, shift registers, and multiplexers/demultiplexers. The following discusses various applications for transmission gates.

Switch Functions

As shown in Figure 8–18, transmission gates can be easily connected to perform the equivalent of the following four common switch functions: SPST, SPDT, DPST, and DPDT. The single-throw (SPST, DPST) functions require no additional logic elements, while the double-throw (SPDT, DPDT) functions require a single 4069 CMOS inverter.

Logic Functions

As was pointed out in Chapter 2, it is possible to construct virtually any type of logic function, given enough NAND gates. The same is almost true of transmission gates. Figure 8–19 shows how two transmission gates are wired to form a 2-input OR gate, identical to a 7432, 4071, or 74C32. Additional inputs may be added by connecting the control input of additional gates at logic 1 and connecting the outputs together. It should be noted that transmission gates may be used in a wired-OR configuration, which is not possible with any other CMOS device.

AND gates are also possible, as shown by the 2-input AND gate in Figure 8–20. This arrangement is equivalent to either a 7408, 4081, or 74C08. When used in conjunction with four pairs of 2-input NOR gates and inverters, a master–slave D-type is formed, as shown in Figure 8–21, and is identical to the 4013. Figure 8–22 illustrates the use of transmission gates in the formation of a J–K flip-flop (identical to a 4027).

Transmission Gates and Operational Amplifiers

Because transmission gates are able to accept analog as well digital signals, they lend themselves well to a number of circuits using operational amplifiers. As shown in Figure 8–23, a series of transmission gates is used to control the gain of an inverting amplifier circuit. If the

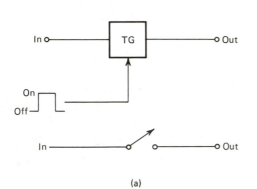

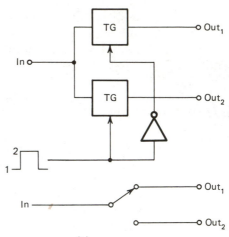

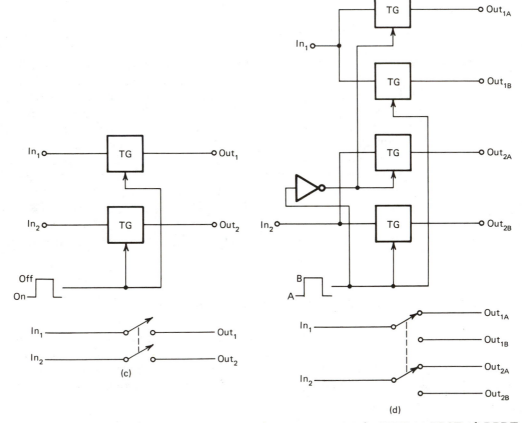

Fig. 8-18 Switch functions from transmission gates: (a) SPST, (b) SPDT, (c) DPST, (d) DPDT.

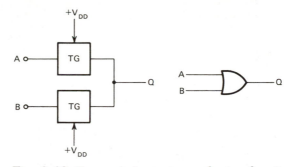

Fig. 8-19 Transmission-gate synthesis of a 2-input OR gate.

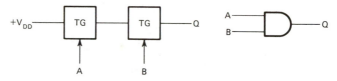

Fig. 8-20 Transmission-gate synthesis of a 2-input AND gate.

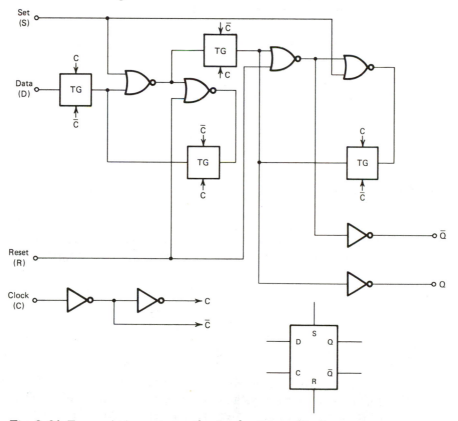

Fig. 8-21 Transmission-gate synthesis of a *D*-type flip-flop.

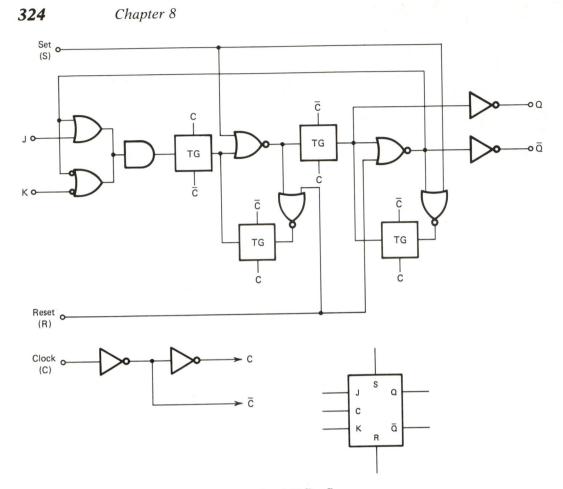

Fig. 8-22 Transmission-gate synthesis of a *J–K* flip-flop.

feedback resistances R_F are at least 10 kΩ, the ON resistance of the transmission gate will have negligible effect on the gain, so

$$\text{Gain} = \frac{V_o}{V_i}$$

$$= -\frac{R_F}{R_I} \qquad (8\text{–}8)$$

In this application, the operational amplifier is usually powered by dual power supplies, as are the transmission gates (e.g., −7.5 and +7.5 V). If the transmission gates are to be enabled by *digital* signals, such as

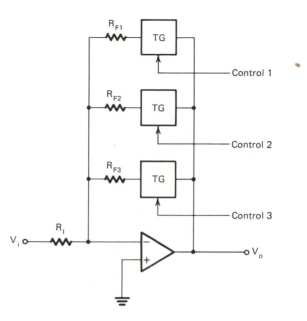

Fig. 8-23 Digital control of op-amp closed-loop gain.

from a computer, operating between ground (logic 0) and V_{DD} (logic 1), the circuit of Figure 8–24 is required for each control input for proper logic-level translation.

Figure 8–25 illustrates how three transmission gates, enabled by a digital signal, are used to invert an analog signal. Gate C is used as a simple inverter, while the control input of the unused gate of a 4016 or 4066 quad package is tied to V_{SS} in order to prevent possible damage to the entire package resulting from high power dissipation.

For active filter sections, transmission gates can be used to shift the center or cutoff frequency instead of using expensive multielement potentiometers or transconductance amplifiers acting as variable resistors. Figure 8–26 illustrates how two transmission gates are used with a *second-order equal-component value* or *Sallen–Key* low-pass filter. The transmission gate is placed in series with each frequency determining resistor R, which should be at least ten times the ON resistance of the gate. If the common control input is a logic 1, both transmission gates are closed, so the cutoff frequency is given by

$$f = \frac{1}{2\pi RC} \qquad (8\text{--}9)$$

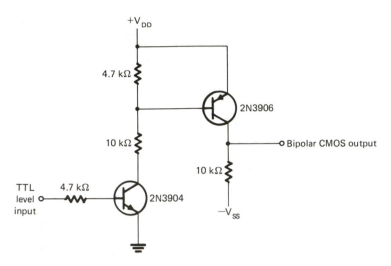

Fig. 8-24 Circuit for TTL to bipolar CMOS voltage level translation.

Now, if the control input is switched ON and OFF between $+V_{DD}$ and $-V_{SS}$ at, for example, a 50:50 rate (50% duty cycle), then the *effective* or *average frequency determining resistance* is *twice* that if the gates were closed all the time. Consequently, the cutoff frequency is now *half* the

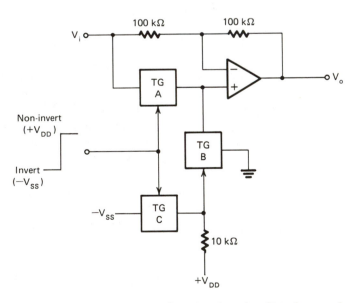

Fig. 8-25 Inverting an analog signal under digital control.

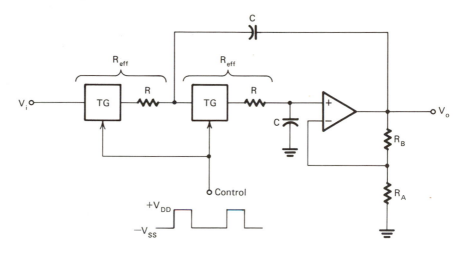

Fig. 8-26 Controlling the cutoff frequency of a second-order low-pass active filter.

former value. Then, by varying the duty cycle of the control pulse train, we can, in turn, shift the cutoff frequency of the filter so that

$$f = \frac{1}{2\pi R_{eff} C} \qquad (8-10)$$

where

$$R_{eff} = \frac{1}{\% \text{ duty cycle}} \times 100 \qquad (8-11)$$

The frequency of the control pulse train must be at least ten times the expected highest cutoff frequency for proper operation.

SCHMITT TRIGGERS

Very often, digital logic must be interfaced with those signals that do not have very sharp rise and fall times. To "square up" a given signal, a *Schmitt trigger* is normally used, which is a *threshold-controlled* switch.

Figure 8–27 illustrates the action of the Schmitt trigger. When the input signal, such as a sine wave, goes above a threshold level, called the

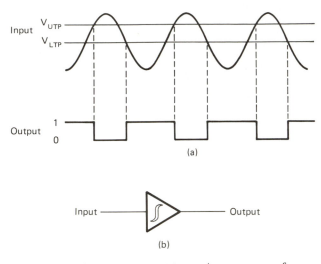

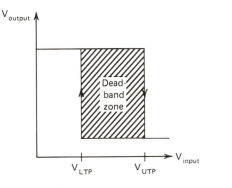

Fig. 8-27 Schmitt trigger: (a) input/output wave forms, (b) schematic symbol.

upper trip point, the output of the Schmitt trigger goes immediately to a logic 0. If the input signal then goes below another threshold level, called the *lower trip point,* the output goes to logic 1. As shown by the Schmitt trigger characteristic curve in Figure 8–28, called a *hysteresis* curve, the difference between these two thresholds is termed the *hysteresis voltage* or *dead-band zone.*

This useful feature allows us to condition certain signals for use with either TTL or CMOS devices. As an example, a 60-Hz (or 50-Hz) pulse train is easily obtained from the sine-wave power-line voltage by

Fig. 8-28 Schmitt trigger hysteresis curve showing the dead-band zone.

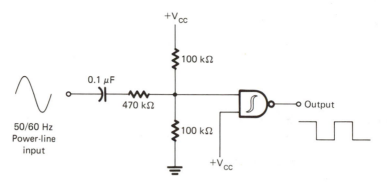

Fig. 8-29 Schmitt trigger interface for obtaining 60-Hz TTL level square waves from the power line frequency.

using the Schmitt trigger circuit of Figure 8–29. The available TTL Schmitt triggers are the 7413, 7414, 74132, and 74232, while CMOS versions are the 74C14, 4093, 40106, and MC14584. Table 8–4 summarizes the upper and lower trip points for these devices.

REVIEW

Although TTL and CMOS devices have different input and output characteristics, this chapter discussed several approaches used to interface these two logic families so that they are compatible. In addition, it was shown how two or more signals or systems share common paths using either open-collector, three-state, or CMOS transmission gate devices. Finally, using a Schmitt trigger, it is possible to interface either TTL or CMOS devices with those signals that do not exhibit sharp rise and fall times.

INTRODUCTION TO THE EXPERIMENTS

The following experiments demonstrate the operation of three-state buffers to create simple bus systems like those used by computers. The experiments that you will perform can be summarized as follows:

Experiment	Purpose
8–1	Create a 1-wire bus that handles four output signals using a 74126 three-state buffer.

Experiment	Purpose
8–2	Create a 4-wire bus system that, in effect, multiplexes the BCD outputs of both a 7490 and a 7493 counter.

EXPERIMENT 8–1

PURPOSE

The purpose of this experiment is to create a 1-wire bus that handles four output signals using a 74126 three-state buffer.

Table 8-4
TTL and CMOS Schmitt Triggers

Device	Type	Number per Package	Supply Voltage	Lower Threshold, V	Upper Threshold, V
7413	4-Input NAND	2	5	0.9	1.7
7414	Inverting	6	5	0.9	1.7
74132	2-Input NAND	4	5	0.9	1.7
74232	2-Input NOR	4	5	0.9	1.7
74C14 (National)	Inverting	6	5	1.4	3.6
			10	3.2	6.8
4093 (RCA)	2-Input NAND	4	5	2.3	2.9
			10	3.9	5.9
40106 (RCA)	Inverting	6	5	1.9	2.9
			10	3.9	5.9
40106 (National)	Inverting	6	5	1.4	3.6
			10	3.2	6.8
MC14584 (Motorola)	Inverting	6	5	2.3	2.9
			10	3.9	5.9

PIN CONFIGURATION OF INTEGRATED-CIRCUIT DEVICE

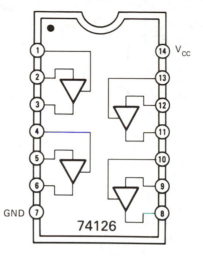

Fig. 8–30 74126 pin diagram.

Required Parts and Equipment

1 74126 TTL quad three-state noninverting buffer
1 LED monitor
4 TTL logic switches
1 TTL logic pulser
1 square-wave clock (approximately 1 Hz)
1 solderless breadboard
1 TTL power supply

SCHEMATIC DIAGRAM OF CIRCUIT

STEP 1

Wire the circuit shown in the schematic diagram of Figure 8–31. When using a three-state device such as the 74126, it is very easy to destroy it if more than one of the ENABLE/DISABLE inputs is at logic 1 at any given time when those corresponding buffer outputs are tied together.

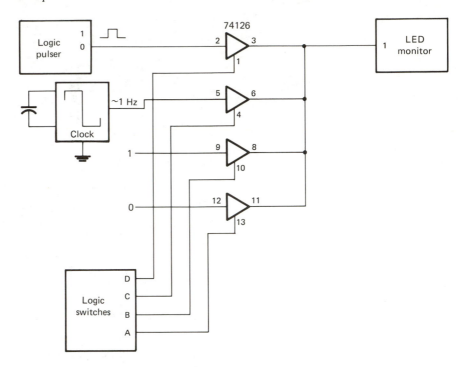

Fig. 8-31 Schematic diagram for Experiment 8–1.

STEP 2

First set *all* logic switches controlling the ENABLE/DISABLE inputs to the four buffer sections to logic 0. Then apply power to the breadboard.

STEP 3

At this point the LED monitor should not be lit. All four buffers are presently *disabled* as all the ENABLE/DISABLE inputs are 0. At this point the output of the buffer essentially looks like an open circuit. Although the LED monitor is unlit, it is behaving as though *nothing* is connected to it.

Now set logic switch *D* at logic 1. The LED monitor should still be off until you activate the pulser. Since the 74126 has *noninverting* buffer sections, the LED monitor now will indicate the logic level of the pulser signal.

STEP 4

Return logic switch *D* to logic 0. Now set logic switch *C* to logic 1. What do you observe on the LED monitor?

You should see the LED flash on and off, which represents the 1-Hz clock signal because only that buffer controlled by logic switch *C* is now enabled.

STEP 5

Return logic switch *C* to logic 0. Now set logic switch *B* to logic 1. What do you observe on the LED monitor?

You should see the LED monitor continuously on, as that buffer controlled by logic switch *B* is now enabled, and its input signal is continuously at logic 1.

STEP 6

Finally, return logic switch *B* to logic 0. Now set logic switch *A* to logic 1. Now what do you observe on the LED monitor?

The LED monitor should be unlit, which represents the logic 0 input signal at the input of that enabled buffer controlled by logic switch *A*.

This arrangement of four three-state buffers is identical to the operation of a 4-to-1 line multiplexer. Depending on which buffer is enabled at any given time, we are then able to pass on that input signal to the output of the buffer, which we have termed a *bus*.

EXPERIMENT 8-2

PURPOSE

The purpose of this experiment is to create a 4-wire bus system that, in effect, multiplexes the BCD outputs of both a 7490 and a 7493 counter.

PIN CONFIGURATION OF
INTEGRATED-CIRCUIT DEVICES

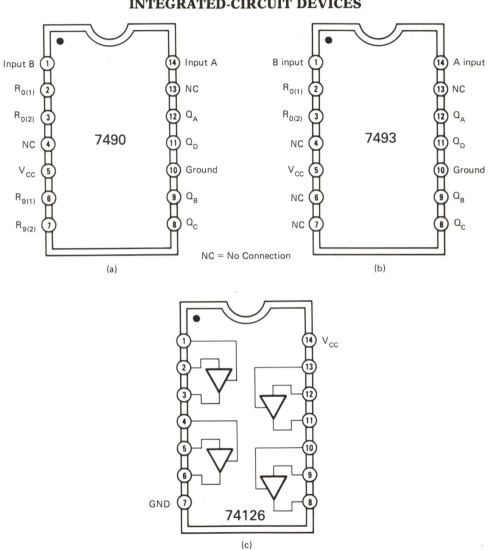

Fig. 8-32 7490, 7493, and 74126 pin diagrams.

Required Parts and Equipment

1   7490 TTL decade counter
1   7493 TTL 4-bit binary counter
2   74126 TTL quad three-state noninverting buffers

1   seven-segment LED display with decoder/driver module
2   TTL logic pulsers
1   solderless breadboard
1   TTL power supply

SCHEMATIC DIAGRAM OF CIRCUIT

STEP 1

Wire the circuit shown in the schematic diagram of Figure 8–33. You should notice that pulser 2 controls both sets of four three-state buffers. However, only one set of four buffers is enabled at any given time to prevent the three-state devices from being destroyed.

STEP 2

Apply power to the breadboard. Which counter's output is currently displayed by the LED display?

Since the four buffers connected to the output of the 7490 counter are enabled, the LED display indicates the output count of the *7490* counter. The outputs of the 7493 counter are effectively disconnected from the 4-wire bus system. Depending on the initial state of the counter when power is first applied, any number between 0 and 9 may appear on the LED display.

STEP 3

Activate pulser 1 several times to clock the counters. You should see the number on the display increase by one each time pulser 1 is pressed and released. Keep activating the pulser until the number 5 is displayed.

STEP 4

Now hold in pulser 2. What do you see on the LED display?

Since the buffers connected to the outputs of the 7490 counter are now disabled, the LED displays the output count of the 7493 (divide-by-16) counter. Depending on the state of the counter when power is first applied, it is possible that any number from 0 to 9, funny looking characters representing the numbers 10, 11,

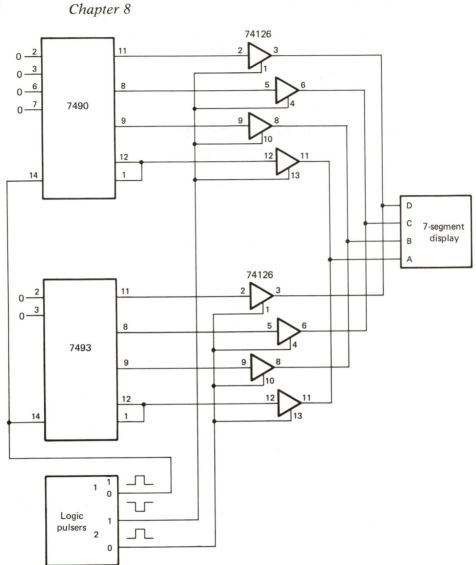

Fig. 8-33 Schematic diagram for Experiment 8–2.

12, 13, and 14, or a blank display representing the equivalent of the number 15 may be displayed. Chances are that you will observe a number that is different than 5 (the present output of the 7490 counter in step 3). Just to make sure, release pulser 2. You should observe the number 5 on the LED display.

STEP 5

Again hold in pulser 2 and press in pulser 1 three times. The number on the LED display should increase by 3. Now release pulser 2. What do you now observe on the LED display?

When you released pulser 2, the number of the LED display changed from whatever the count on the 7493 counter was to 8, which is again the output of the 7490 counter. In this experiment, both counters are clocked simultaneously. By pressing pulser 1 in step 5, you have also clocked the 7490 counter three times, thereby increasing its displayed count from 5 to 8. Although both counters are being clocked together, the LED display shows the output of only one of the two counters at any given time.

This arrangement of eight three-state buffers is identical to the operation of four 2-to-1 line multiplexers. Depending on which four buffers are enabled at any given time, we are then able to pass the BCD outputs of either the 7490 or the 7493 counter on to a 4-wire bus system.

9

INTEGRATED-CIRCUIT MEMORIES

INTRODUCTION

At this point, we have shown several examples of how to combine various logic functions to perform specific tasks. However, if we want to change the task to another, we may have to physically rewire the circuit. On the other hand, the heart of today's computer is built around a *microprocessor*, which has the ability to be reprogrammed so that one can easily change the task to another.

For a computer to function, it must have some form of memory. For example, a microprocessor-controlled microwave oven may at one time be told to defrost meat for 10 minutes and then cook it at 350 degrees for 30 minutes. On another occasion, it may be used to cook a slice of pizza or a hot dog without prior defrosting for only 2 minutes. At the completion of both tasks, a bell is sounded to tell you that the total cooking cycle has been completed. For the microwave oven to "remember" any sequence of commands entered on the oven's front panel for a given cooking task, it must have some form of programmable memory to remind the microprocessor "brain" what to do next.

The area of integrated-circuit memories is so extensive that an entire book could be specifically devoted to it. Taking this into account, this chapter presents only the basic concepts of the major types of integrated-circuit memories that are in use.

339

OBJECTIVES

After completing this chapter, you will be able to:

- *Explain the difference between bit- and word-organized memories.*

- *Explain the difference between RAM and ROM.*

- *Understand write and read cycle timing.*

- *Explain the basic input and output connections required for either RAMs or ROMs.*

- *Describe several schemes for expanding memory capacity using both bit and word organized memories.*

- *Explain the difference between static and dynamic RAMs.*

WHAT ARE SEMICONDUCTOR MEMORIES?

Integrated-circuit memories are digital devices that are capable of storing logic 0s and 1s in a predefined arrangement. In a very loose sense, flip-flops and shift registers can be thought of as "memories" since their outputs will remain in a given state even after the input data are removed and will retain these states until changed (reset, application of a clock pulse, etc.).

As shown in the family tree diagram of Figure 9-1, integrated-circuit memories can be divided into two types: read-only and random

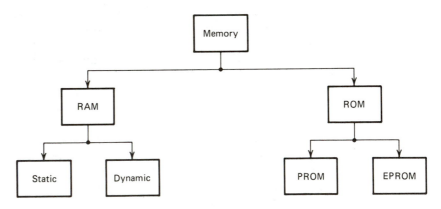

Fig. 9-1 Family tree diagram of memory types.

access, each of which can be further subdivided. Each classification is discussed in the following sections.

Memory Organizations

Before discussing the characteristics of the various memory types, let's first discuss the basic characteristics that are common to all. As shown in Figure 9–2, a typical memory device can be thought of as being organized in terms of a matrix arrangement. Each matrix location, or *cell*, is selected by a row and column location. This cell location is selected by an *address decoder*. As a general scheme, an internal decoder selects the row of the particular memory cell, while another decoder selects the corresponding column of the cell. The binary inputs to the decoders are referred to as the *memory address*.

As far as the internal memory is organized, there are two conventions: bit organized and word organized. A *bit-organized* memory device has the capability to store only a single bit for each address location. Consequently, this arrangement has a single data input line, and a single but usually separate data output line, as shown in Figure 9–3a. However, some bit-organized memory devices have a combined input/output (I/O) data line with a separate I/O control line that selects whether this combined data line functions as the input or output line at a given time (Figure 9–3b).

On the other hand, in a *word-organized* memory device, a memory address selects *several* memory cells at one time instead of

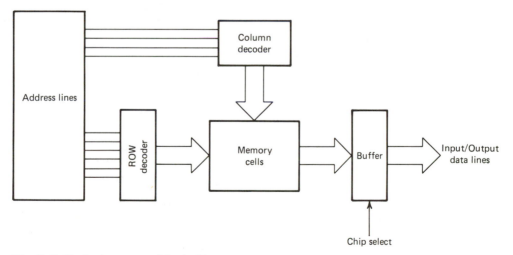

Fig. 9-2 Typical memory block diagram.

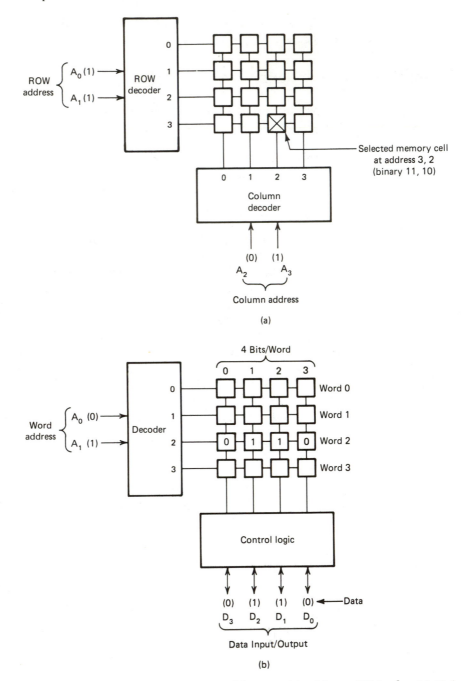

Fig. 9-3 Decoding binary memory addresses: (a) address 1011 of a 16 × 1 bit-organized memory, (b) address 10 of a 4 × 4 bit word-organized memory.

only a single cell. The number of cells selected simultaneously is the *word length* and is usually 4 or 8 bits in length. Also, word-organized memories have the same number of I/O lines as there are bits in the word.

The 1K Convention

Whether bit or word organized, integrated-circuit memories are further characterized as to storage capacity. For memory capacities larger than 1024 bits, the suffix K is generally used as a shorthand notation. Table 9–1 illustrates this shorthand notation.

Since 1K equals a total capacity of 1024 bits, a 4K memory then has 4×1024, or 4096 bits. It should be noticed from Table 9–1 that 65,536 bits is abbreviated as 64K, although some sources incorrectly use 65K. The same inconsistency carries through for capacities greater than 64K.

As bit-organized memories have a word length of 1 bit, they are generally characterized as $nK \times 1$ memories. As an example, a $16K \times 1$ (bit organized) memory has a total capacity of 16K words $\times$ 1 bit/word, or 16K bits (cells) of memory. In actuality, there are really 16,384 cells. On the other hand, word-organized memories, as explained earlier, generally have word lengths of 4 or 8 bits. Consequently, a $4K \times 4$ (word organized) memory device has 4K (4096) words of 4 bits each for a total of 16K bits. As another possibility, a memory device may have a $2K \times 8$ organization, which still has 16K bits of storage. Whether $16K \times 1$, $4K \times 4$, or $2K \times 8$, all three arrangements have the same storage capacity: 16K or 16,384 bits.

Table 9–1

Bits	Equivalent Notation
1,024	1K
2,048	2K
4,096	4K
8,192	8K
16,384	16K
32,768	32K
65,536	64K

For memory segments smaller than 1K, the term *page* is frequently used to indicate the equivalent of 256 bits, so that 4 pages equals 1K. For a 64K memory system, there are 4×64 or 256 pages.

Input/Output Connections

Besides the necessary power supply requirements, connections to and from integrated-circuit memories can be divided into the following three categories: address inputs, data input/output, and control inputs.

1. Address Inputs

All memories need a series of *address inputs* to select the proper memory cell, or cells in the case of a word-organized device. The total number of address inputs or *lines* depends on the total number of addressable words. If a memory device has a storage capacity of X words, then it requires N address lines, so

$$N = \frac{\log X}{\log 2} \qquad (9\text{--}1)$$

As an example, a $16K \times 1$ memory requires 14 address lines, while a $2K \times 8$ device requires 11 address lines. Even though each device has a total capacity of 16K bits, the first example has 16K words while the second example has only 2K words.

When looking at schematic diagrams containing integrated-circuit memories, address lines are usually denoted by the letter A, so that the 11 address lines for a $2K \times 8$ device would be numbered in binary from A_0 to A_{10}. To specify the address of the 547th 8-bit cell, the 11 binary memory address lines would have the form

$$\begin{array}{ccccccccccc} 0 & 1 & 0 & 0 & 0 & 1 & 0 & 0 & 0 & 1 & 1 \end{array}$$
$$\quad A_{10} \hspace{6.5cm} A_0$$

which can be verified by

$$2^9 + 2^5 + 2^1 + 2^0 = 512 + 32 + 2 + 1 = 547$$

2. Data I/O Lines

The number of equal *data I/O lines* depends on the word length of the memory device. A $4K \times 1$ device will have one data input and one output line, while a 1K × 8 device will have 8 data I/O line pairs. There may be separate sets of lines for input and output, or there may be only one set of lines that handles both input and output data. Such a dual-purpose arrangement is generally referred to as *bidirectional* and is generally used in computer systems to reduce the number of required connections.

When looking at schematic diagrams containing integrated-circuit memories, the data I/O lines are usually denoted by the letter D, so the 8 data lines for a 2K × 8 device would be numbered in binary from D_0 to D_7. Consequently, the 8-bit binary word 10011101 would appear on the 8 data I/O lines as follows:

$$1 \quad 0 \quad 0 \quad 1 \quad 1 \quad 1 \quad 0 \quad 1$$
$$D_7 \hspace{5.5cm} D_0$$

3. Chip Select Input

The *CHIP SELECT ($\overline{CS}$)* or *CHIP ENABLE ($\overline{CE}$)* input is one of several available control inputs. This input is used to select, activate, or enable the memory device before data can be written to or read from it. As a general rule, integrated-circuit memories have three-state data outputs and, when used with other memory devices to increase the available memory capacity, the $\overline{CS}$ (or $\overline{CE}$) control input is used to effectively isolate that device from the rest of the circuit. Usually, a logic 0 is required to enable or select a given memory device; a logic 1 disables it. Techniques using the CHIP SELECT input to expand memory capacity will be discussed later in this chapter. Other control inputs that apply to ROMs or RAMs are also discussed in later sections where appropriate.

Address Timing

Integrated-circuit memories, in addition to their storage capacity, are rated as to their speed. In general, there are three "speed" parameters that are important to know when specifying integrated-circuit memories. They are

- Address access time
- Chip select access time
- Memory read cycle time

The *address access time* is the time period that it takes for the memory device to output the data stored at a particular address after that particular address has been selected. The majority of integrated-circuit memories have address access times that fall between 450 ns and 150 ns, with the 150-ns access time characterizing the "faster" device. However, you get what you pay for. Memory devices that have faster access times will be more expensive.

The *chip select access time* is the time period, after the memory's CHIP SELECT input is activated, that the memory device is capable of outputting its data and is generally shorter than the address access time. This is because the memory's address is usually already selected before the CHIP SELECT input is activated. The *memory read cycle time* is the time period for the memory to complete reading its data and then become available for the next cycle, which may be either read or write.

READ-ONLY MEMORY

The *read-only memory*, or ROM, is a type of memory device whose contents, once programmed, cannot be easily changed. A ROM is sometimes referred to as a *nonvolatile* memory because it will retain its stored information even if the power supply voltage to it is removed.

As shown by the family tree of Figure 9–1, ROMs come in several categories, depending on whether or not their memory contents can be erased, or cleared. Some ROMs, once programmed, do not have the capability to have their contents changed under any circumstances. These type are said to have their contents "burned in." Other types of ROMs do permit their contents to be changed, but this is an all or none proposition. That is, *all* memory contents are erased or *zeroed out*.

A *programmable read-only memory*, or PROM, internally contains cell circuits with *fuses*. When each cell is programmed to either a 0 or a 1, its "fuse is blown" so that its contents can never be erased or changed. On the other hand, ROMs such as *EPROMs*, or *erasable programmable read-only memories*, once programmed can have all their cells erased by exposing them to an ultraviolet light source of sufficient intensity for a sufficient amount of time. Unlike PROMs, there is either a circular or square-shaped quartz window on the top surface of an EPROM for the ultraviolet light to shine through, as illustrated in Figure 9–4. Once programmed, EPROMs should have this window covered with a piece of dark tape to prevent accidental erasure of their contents.

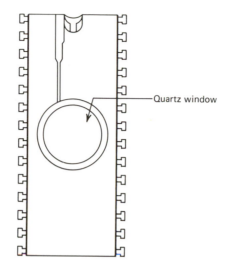

Fig. 9-4 Top view of an EPROM showing the quartz window.

Applications Using Read-Only Memories

Since the stored data are essentially permanent, ROMs are generally used when a given sequence of bits or grouped data are required many times without changes. The following examples illustrate how ROMs are principally used.

1. Code Converters

ROMs can be used to change from one type of code to another. Suppose we need to convert the sixteen 4-bit combinations of Table 9–2 to their binary equivalents. For example, every time we encounter the 4-bit sequence 0100, we want to translate or convert this to the binary sequence 0111, representing the decimal number 7. If we use a 64-bit ROM, organized as 16×4-bit words, the ROM can be permanently programmed to convert any one of these sixteen 4-bit combinations into its equivalent 4-bit binary code. Since this ROM example is *word organized*, there are four address or input lines and four output data lines. This is an example of one of many types of code conversions that ROMs can perform.

In Chapters 6 and 7 we discussed several types of code converters using both combinatorial logic as well as specific TTL and CMOS devices. For some applications, a ROM can be programmed to function

Table 9-2

Input				Output			
A_3	A_2	A_1	A_0	Q_3	Q_2	Q_1	Q_0
0	0	0	0	0	0	0	0
0	0	0	1	0	0	0	1
0	0	1	1	0	0	1	0
0	0	1	0	0	0	1	1
0	1	1	0	0	1	0	0
0	1	1	1	0	1	0	1
0	1	0	1	0	1	1	0
0	1	0	0	0	1	1	1
1	1	0	0	1	0	0	0
1	1	0	1	1	0	0	1
1	1	1	1	1	0	1	0
1	1	1	0	1	0	1	1
1	0	1	0	1	1	0	0
1	0	1	1	1	1	0	1
1	0	0	1	1	1	1	0
1	0	0	0	1	1	1	1

identically as those converters of earlier chapters. In spite of this feature, however, ROMs are not often used in place of most TTL and CMOS decoders, as the ROM would generally cost more to implement.

2. Character Generators

For a computer to display characters on a cathode-ray tube, or video display, the computer must have the ability to tell the display how to form a variety of letters, numbers, and special symbols. In Chapter 7, we discussed using a 5 × 7 dot matrix LED display, where specific LED *dots* must be lit to form the capital letter G. Since there are 35 separate possible LED dots, arranged in a 5-column by 7-row format, it is possible to create literally hundreds of different standard and nonstandard characters by merely selecting the desired dots.

3. Computers

Computers frequently use ROMs to store a sequence of instructions of a computer program. As an example, most microcomputers are 8-bit types, although 16-bit systems are rapidly becoming commonplace. That is, each basic computer instruction is either 8 or 16 bits in length. With an 8-bit computer system, for example, it is possible to address 64K or 65,536 different memory locations of 8 bits each. The total possible storage capacity is then 64K $\times$ 8 or 512K (524,288) bits.

In an 8-bit computer system, ROM usually occupies a small portion of the total 64K memory available, typically less than 16K (16,384) 8-bit words. As an example, suppose we wish to have a 1K $\times$ 8 ROM system, which contains one or more permanently stored programs. With all the different memory types available, both bit and word organized, there are many possible ways to accomplish this arrangement. The easiest is to simply use one of the many 1K $\times$ 8 ROMs that are available, as shown by the ROM circuit of Figure 9–5.

On the other hand, the same required 1K $\times$ 8 organization can be accomplished by using, for example, two 512 bit $\times$ 8 ROMs, illustrated by the circuit of Figure 9–6. In this arrangement, the basic 512 bit $\times$ 8 bit device configuration has been expanded to a 1K $\times$ 8 system by connecting the eight three-state output data lines D_0 through D_7 of both devices in parallel. According to Eq. 9–1, ten address lines (A_0 through

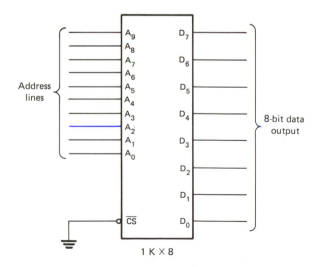

Fig. 9-5 Schematic diagram for a 1K $\times$ 8 ROM device.

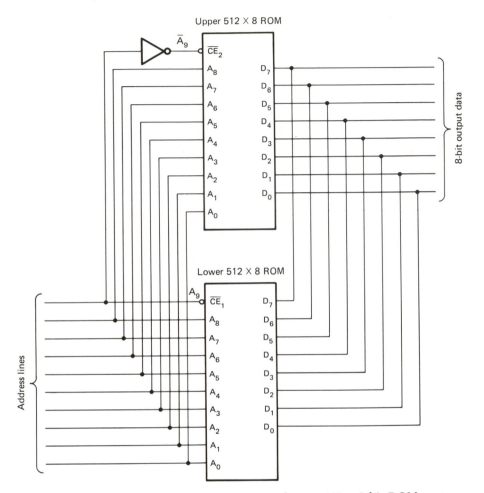

Fig. 9-6 Combining two 512 × 8 bit ROMs to form a 1K × 8 bit ROM system.

A_9) will be required to access any one of 1024 possible locations. However, each 512 bit × 8 ROM has only *nine* such lines ($2^9 = 512$), A_0 through A_8, which are connected in parallel. The tenth address line (A_9), which is the most significant address bit, is connected in complementary fashion to the CHIP SELECT inputs of both ROMs.

For the "lower" 512 bit × 8 ROM, the CHIP SELECT input ($\overline{CE_1}$) is connected to the most significant address line (A_9), while the CHIP SELECT input ($\overline{CE_2}$) of the "upper" 512 bit × 8 ROM is connected to its complement ($\overline{A_9}$). As long as A_9 is 0, only the lower 512 bit × 8 ROM is enabled, so the maximum 10-bit binary address possible is 0111111111, which is equivalent to 511 and says that there are *512* possible 8-bit memory locations, counting the address 0000000000 as a valid address. Once the lower or first 512 8-bit locations have been addressed, address line A_9 must now be a 1 in order to address those locations greater than 512, which then disables the "lower" ROM, while enabling the "upper" ROM, which can also store 512 8-bit locations. Consequently, we are then able to select either ROM, giving a total capacity of 1,024 eight-bit locations, although only *512* eight-bit locations are available at any given time.

Another possible scheme, shown in Figure 9–7, is to organize a 1K $\times$ 8 ROM system by using eight 1K $\times$ 1 ROMs. This time, using bit-organized ROMs, the ten address lines of all eight ROMs are connected in parallel, while all $\overline{CS}$ inputs are permanently connected to ground. Consequently, all eight ROMs are enabled simultaneously.

RANDOM-ACCESS MEMORY

The *random-access memory* or RAM gets its name from the fact that any given memory cell (address) can be selected in any sequence (i.e, at random). Furthermore, RAMs have the ability to have the data (contents) of any memory address changed at any time. This is referred to as a *write* operation. Like a ROM, a RAM also has the ability to output (i.e., *read*) the data stored at a given address. Because of this twofold function, RAMs are referred to as *read/write memories*. However, unlike ROMs, RAMs are also classified as *volatile memories*, as a RAM will lose its data contents if the power to the device is briefly removed.

RAMs are generally divided into two major categories: static and dynamic. A *static RAM* retains its data indefinitely as long as power is supplied to it. On the other hand, the data stored by a *dynamic RAM* must be periodically *refreshed* or updated. This is usually accomplished by performing either a repetitive read or write operation. Despite having to be periodically refreshed, dynamic RAMs are available with storage capacities larger than those of static RAMs.

RAMs can be expanded in the same way that ROMs are (see Figures 9–5 through 9–7). Since RAMs are able to be written to as well as read from, a read/write ($\overline{R/W}$), or write enable ($\overline{WE}$) line is used to select either operation. For *static* RAMs, this is the only extra control input.

As an example, Figure 9–8 shows a 16K $\times$ 1 dynamic RAM, a type that is found in many of today's microcomputers. For a 16K device, there must then be 14 address lines, which can be organized in a 7-row, 7-column matrix. The 1-bit data word is stored in the memory cell

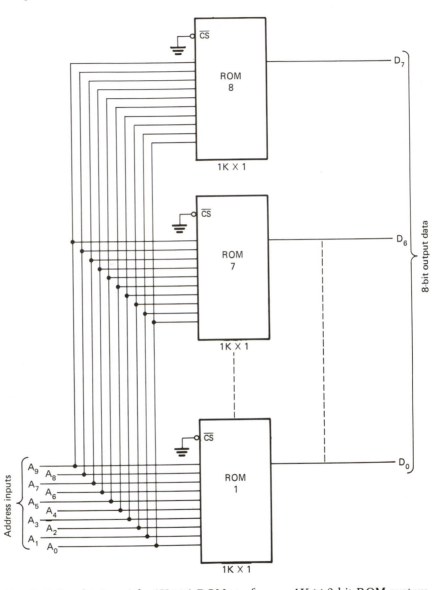

Fig. 9-7 Combining eight 1K × 1 ROMs to form a 1K × 8 bit ROM system.

selected partially by the 7-row address bits, A_0 to A_6, when a *row address strobe ($\overline{RAS}$)* signal is at logic 0. The remainder of the 14-bit address is comprised of the 7-column address bits, A_7 to A_{13}, and is selected when the *column address strobe ($\overline{CAS}$)* signal is also at logic 0. Since the RAM is organized as 16K 1-bit words, there is only a single

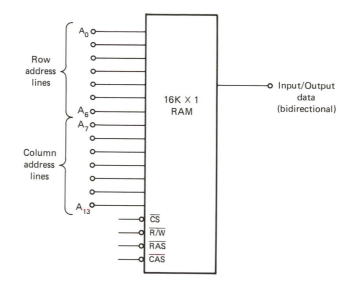

Fig. 9-8 Schematic diagram for a 16K × 1 RAM device.

bidirectional data I/O line, which is frequently referred to as an *I/O bus*. Depending on the logic level of the $\overline{R/W}$ input at a given time, the data I/O line either receives this 1-bit data and *writes* it into one of 16,384 possible memory cells or outputs it (i.e, *reads* it) over the same line.

Since these types of 16K RAMs are widely used in microcomputer systems, such as the 8-bit Apple-II and Radio Shack's TRS-80 systems, there must then be eight 16K RAMs, as shown in a block diagram form by Figure 9-9. The 14 address lines are connected in parallel as an address bus. The refresh lines, $\overline{RAS}$, $\overline{CAS}$, as well as the $\overline{CS}$, and $\overline{R/W}$ lines are also connected in parallel.

REVIEW

In this chapter we have taken a little more than a cursory look at integrated-circuit memories, but, in keeping with the intended scope of this book, discussed most of the concepts in general terms. The major types of integrated-circuit memories were divided into read-only (nonvolatile) and random-access (volatile) types. We also learned that memories can be either bit or word organized and that storage capacity can be expanded in a variety of ways.

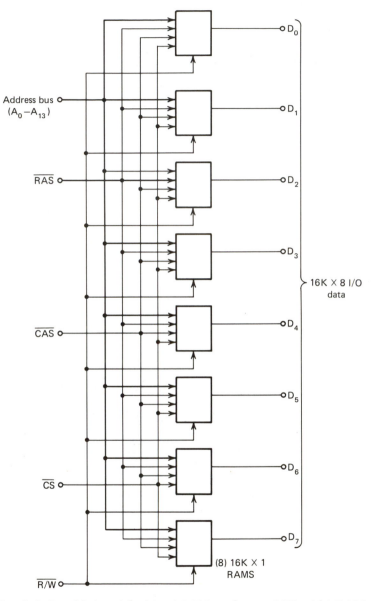

Fig. 9-9 Combining eight 16×1 RAMs to form a $16K \times 8$ bit RAM system.

INTRODUCTION TO THE EXPERIMENTS

The following two experiments are intended to demonstrate some of the basic characteristics of the two broad classes of semiconductor memories: read only and random access. The experiments that you will perform can be summarized as follows:

Experiment	Purpose
9–1	Demonstrates the operation of a 7447 decoder as a nonvolatile ROM.
9–2	Demonstrates how to write data to and read data from a 7489 TTL 16 × 4-bit RAM.

EXPERIMENT 9–1

PURPOSE

The purpose of this experiment is to demonstrate the operation of a 7447 TTL decoder as a nonvolatile 16-word by 7-bit read-only memory.

PIN CONFIGURATION OF
INTEGRATED-CIRCUIT DEVICE

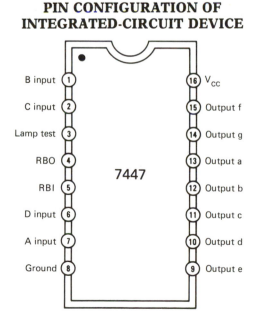

	7447	
B input (1)		(16) V_{CC}
C input (2)		(15) Output f
Lamp test (3)		(14) Output g
RBO (4)		(13) Output a
RBI (5)		(12) Output b
D input (6)		(11) Output c
A input (7)		(10) Output d
Ground (8)		(9) Output e

Fig. 9-10 7447 pin diagram.

1 7447 TTL BCD-to-7-segment decoder/driver
4 TTL logic switches
7 LED monitors
1 solderless breadboard
1 TTL power supply

SCHEMATIC DIAGRAM OF CIRCUIT

STEP 1

Wire the circuit shown in the schematic diagram of Figure 9–11. Apply power to the breadboard and set all four logic switches at logic 0 so that the memory address is *DCBA* = 0000. Which LED monitors, if any, are lit?

You should observe that all the LED monitors are lit except LED monitor G. Record this result in Table 9–3, using 0s and 1s to indicate the state of each of the seven LED monitors.

STEP 2

In turn, set the 4-bit memory address to those listed in Table 9–3 and record your results for each of the 15 remaining memory addresses.

STEP 3

Once you have examined and recorded the data stored in all 16 memory locations, with each data word having a length of 7 bits, you

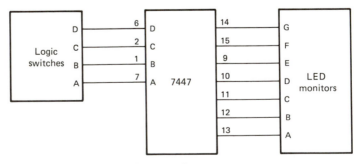

Fig. 9-11 Schematic diagram for Experiment 9–1.

Table 9-3
Data for Experiment 9-1

Memory Address				Memory Output						
D	*C*	*B*	*A*	*A*	*B*	*C*	*D*	*E*	*F*	*G*
0	0	0	0							
0	0	0	1							
0	0	1	0							
0	0	1	1							
0	1	0	0							
0	1	0	1							
0	1	1	0							
0	1	1	1							
1	0	0	0							
1	0	0	1							
1	0	1	0							
1	0	1	1							
1	1	0	0							
1	1	0	1							
1	1	1	0							
1	1	1	1							

should be able to go back to any one of the 16 locations and see that the 7-bit data word for that location remains the same. Try it.

STEP 4

Now remove the power supply connections from the breadboard, wait a few seconds, and again apply power to the breadboard. Repeat steps 1 and 2. Are the results any different from the first time that you performed this experiment?

You should find that outputs representing the stored data remain the same for a given memory address no matter how many times you disconnect and reapply the power to the breadboard. When applied to memories, this characteristic is called *nonvolatile*. Although you have used the 7447 device previously in Chapter 7 as a BCD-to-7-segment LED display decoder/driver, this experiment illustrates how the 7447 also can be thought of as a *nonvolatile ROM* whose permanently stored memory contents cannot be changed. Besides acting as a ROM, the 7447, when connected to a common-anode seven-segment LED display, can also be thought of as a *character generator*. This was demonstrated earlier in Experiment 7–3.

EXPERIMENT 9–2

PURPOSE

The purpose of this experiment is to write data to and read data from a 7489 TTL 16 × 4-bit RAM.

PIN CONFIGURATION OF INTEGRATED-CIRCUIT DEVICES

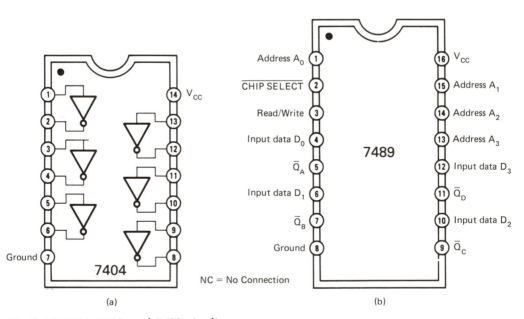

Fig. 9-12 7404, 7489, and 7493 pin diagrams.

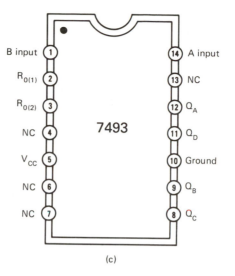

Fig. 9-12 (*Continued*)

Required Parts and Equipment

4 1-kΩ resistors
1 7404 TTL hex inverter
1 7489 TTL 16 × 4 RAM
1 7493 4-bit binary (divide-by-16) counter
4 TTL logic switches
2 TTL pulsers
4 LED monitors
1 seven-segment LED display with decoder/driver
1 solderless breadboard
1 TTL power supply

SCHEMATIC DIAGRAM OF CIRCUIT

STEP 1

Wire the circuit shown in the schematic diagram of Figure 9–13. If you do not have a seven-segment LED display module, use the display circuit of Experiment 7–3. Otherwise, use four additional LED monitors.

The 7489 RAM is *word organized,* so data can be stored as sixteen 4-bit words or a total of 64 bits. Since the four output lines of the 7489

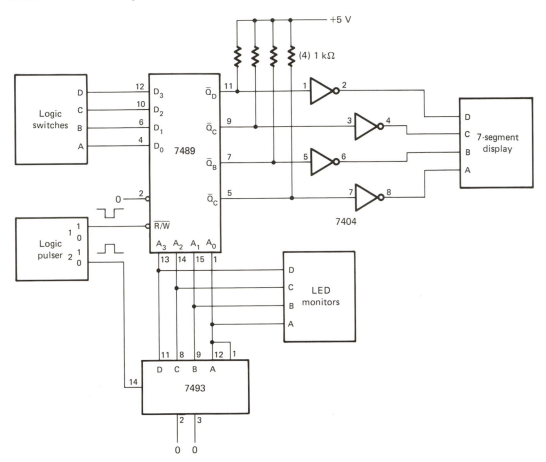

Fig. 9-13 Schematic diagram for Experiment 9–2.

are of the open-collector type, 1-kΩ pull-up resistors are required. Furthermore, these outputs are also *inverted* with respect to the stored data, which is why four sections of a 7404 hex inverter are used to take the complement. Pulser 1 allows data to be loaded or written into memory, while pulser 2 sequences all 16 memory addresses via a 7493 counter.

STEP 2

After carefully checking your wiring, apply power to the breadboard. Initially, the 7493 counter may be in any one of 16 states. Using pulser 2, cycle the 7493 counter so that the memory address of the RAM, as indicated by LED monitors A to D, equals *DCBA* = 0000.

STEP 3

Now cycle through all 16 memory locations and record what appears on the seven-segment LED display in Table 9–4 for each location.

STEP 4

Now set the four logic switches so that the input data are $DCBA = 0101$ (decimal 5). Cycle the 7493 counter so that the RAM address location is $DCBA = 0000$. Press and release pulser 1. What appears on the seven-segment LED display?

You should see the number 5 on the LED display.

Table 9-4
Data for Experiment 9-2

Memory Address				Displayed Symbol
D	C	B	A	
0	0	0	0	
0	0	0	1	
0	0	1	0	
0	0	1	1	
0	1	0	0	
0	1	0	1	
0	1	1	0	
0	1	1	1	
1	0	0	0	
1	0	0	1	
1	0	1	0	
1	0	1	1	
1	1	0	0	
1	1	0	1	
1	1	1	0	
1	1	1	1	

STEP 5

Now cycle the 7493 counter so that the RAM address location is $DCBA = 1011$. Set the logic switches so that the input data are $DCBA = 1000$ (decimal 8). Press and release pulser 1. What appears on the seven-segment LED display?

You should now see the number 8 on the LED display. In both cases you stored or wrote the 4-bit equivalents of the decimal numbers 5 and 8, respectively, into the memory locations whose addresses are $DCBA = 0000$ and 1011.

STEP 6

Starting with memory address $DCBA = 0000$, cycle through all 16 memory locations and compare what is seen on the seven-segment LED display with the observations recorded earlier in Table 9–4. Are there any differences?

You should find that everything remains the same except the data displayed at memory locations $DCBA = 0000$ and 1011. This is because, in steps 4 and 5, you stored new data in these two locations. Each time you press in pulser 1, the read/write (or write enable) input at pin 3 is at logic 0 and the 4-bit data present at pins 4, 6, 10, and 12 (D_0 to D_3) are stored in the memory location determined by the status of the address inputs A_0 to A_3 at pins 1, 15, 14, and 13. When the $\overline{R/W}$ input is 1, the complements of the stored contents of the memory location appear at pins 5, 7, 9, and 11.

STEP 7

Now store the indicated 4-bit data words at the memory locations listed in Table 9–5. Then remove the power from the breadboard, wait a few seconds, and again apply power to the breadboard. Starting at address $DCBA = 0000$, compare the data now displayed with those listed in Table 9–5.

Chances are that none of the data in the 16 memory locations is exactly the same as that you have just stored. In effect, you have just lost the stored data forever! This is the *volatility* nature of a RAM, or *read/write* memory. If power is momentarily lost, the stored contents will be lost. In our case, it was found that the data

Table 9-5
Data for Experiment 9-2

Memory Address				Input Data			
D	C	B	A	D	D	D	D
0	0	0	0	1	0	1	0
0	0	0	1	1	1	0	1
0	0	1	0	1	1	1	1
0	0	1	1	1	0	0	1
0	1	0	0	0	1	0	1
0	1	0	1	1	0	1	1
0	1	1	0	0	1	1	0
0	1	1	1	0	0	0	1
1	0	0	0	0	0	1	1
1	0	0	1	1	0	0	0
1	0	1	0	1	1	1	0
1	0	1	1	0	1	1	1
1	1	0	0	0	1	0	0
1	1	0	1	1	1	0	0
1	1	1	0	0	0	0	0
1	1	1	1	0	0	1	0

present at all 16 memory locations were the same as were recorded in Table 9-4 when we initially applied power to the RAM. This may be an unusual characteristic of our 7489 RAM, but it should be concluded that this device acts as a nonvolatile ROM, since the stored data are the same each time power is applied to the RAM.

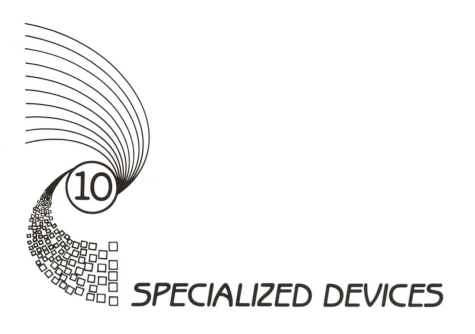

SPECIALIZED DEVICES

INTRODUCTION

Throughout this book, where appropriate, TTL or CMOS gates, latches, counters, and the like, have been shown to be easily connected to create entire systems, such as digital time clocks, frequency/event counters, and digital panel meters to name a few. However, when such systems are built, the total number of individual integrated circuits required can be overwhelming.

Rather than reinvent the wheel each time, several manufacturers have developed a number of useful single-chip systems, which are intended to replace a majority, if not all, the individual devices of a functionally equivalent circuit. This final chapter discusses several of these one-chip systems, such as phase-locked loops and synthesizers, multidigit counters/frequency counters, Touch-Tone encoders, oscillators/dividers, and 3½ digit A/D converters.

OBJECTIVES

After completing this chapter, you will be able to do the following:

- *Learn how to use single-chip systems to reduce the number of components with the following devices as examples:*

- *MC14410 and ICM7206 Touch-Tone encoders.*

365

- *MC14433 and ICL7106/7107 3½ digit A/D converters.*

- *4046 phase-locked loop.*

- *ICM7207/7208 two-chip frequency counter.*

TOUCH-TONE ENCODERS

Stable Touch-Tone* signals used for telephone dialing and other communication services can be encoded by several CMOS devices using readily available quartz crystals. Touch-Tone signals are encoded in *tone-pairs* using two of eight possible tone frequencies for the numbers 0 through 9, letters A through D, and the symbols # (pound) and * (star). The possible eight audio frequencies are shown in matrix form in Table 10–1, where the tones are divided into a low- and a high-frequency group. Representative CMOS encoders are the MC14410 and the ICM7206.

The Motorola MC14410 (Figure 10–1) permits the design of a low-power Touch-Tone encoder using a standard 1-MHz crystal, as shown in Figure 10–2, and features multiple key lockout with a standby power consumption of approximately 15 mW using a 5-V supply.

The inputs from a 2-of-8 key pad originate from a 4×4 switch matrix generating a four row ($R1$ to $R4$) by four column ($C1$ to $C4$) input signal according to Table 10–2. As an example, if the number 7 is pressed on the key pad, the row 3 ($R3$) and column 1 ($C1$) inputs are simultaneously grounded, which in turn programs the MC14410's

Table 10-1
Audio Frequencies Used for Touch-Tone Encoding

Low-Tone Group (Hz)	High-Tone Group (Hz)			
	1209	1336	1477	1633
697	1	2	3	A
770	4	5	6	B
852	7	8	9	C
941	*	0	#	D

---

*Touch-Tone is a registered trademark of American Telephone and Telegraph.

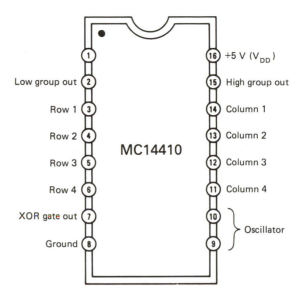

Fig. 10-1 MC14410 CMOS Touch-Tone encoder pin diagram.

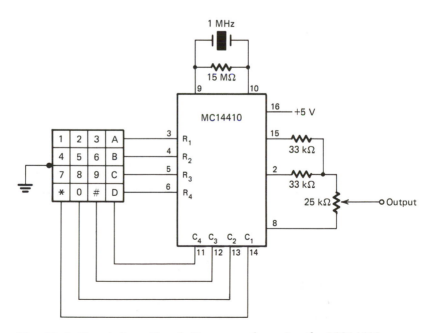

Fig. 10-2 Circuit for a Touch-Tone encoder using the MC14410.

Table 10-2
Touch-Tone Encoder Truth Table

Character Input	Row				Column				High Tone (Hz)	Low Tone (Hz)
	R1	*R2*	*R3*	*R4*	*C1*	*C2*	*C3*	*C4*		
1	0	1	1	1	0	1	1	1	1209	697
2	0	1	1	1	1	0	1	1	1336	697
3	0	1	1	1	1	1	0	1	1477	697
A	0	1	1	1	1	1	1	0	1633	697
4	1	0	1	1	0	1	1	1	1209	770
5	1	0	1	1	1	0	1	1	1336	770
6	1	0	1	1	1	1	0	1	1477	770
B	1	0	1	1	1	1	1	0	1633	770
7	1	1	0	1	0	1	1	1	1209	852
8	1	1	0	1	1	0	1	1	1336	852
9	1	1	0	1	1	1	0	1	1477	852
C	1	1	0	1	1	1	1	0	1633	852
*	1	1	1	0	0	1	1	1	1209	941
0	1	1	1	0	1	0	1	1	1336	941
#	1	1	1	0	1	1	0	1	1477	941
D	1	1	1	0	1	1	1	0	1633	941

internal decoder logic to simultaneously deliver both a 852-Hz low tone and a 1209-Hz high tone. The two tones are then added together via the two 33-kΩ resistors to give the desired two-tone output. The output at pin 7 is a pulsed logic output obtained by feeding the outputs of the internal frequency dividers of the two tone groups to an XOR gate. This output can be used as a keyed "push-to-talk" signal to drive a one-shot when any of the keys are briefly pressed.

The Intersil ICM7206 (Figure 10–3) is somewhat similar to the MC14410. It also consumes less than 8 mW and has multiple-key lockout. However, the 7206 requires a 3.579545-MHz color TV crystal, as shown in the operating circuit of Figure 10–4.

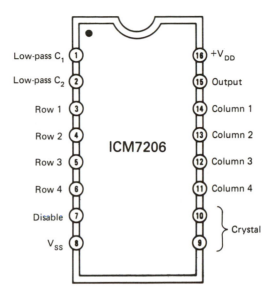

Fig. 10-3 ICM7206 CMOS Touch-Tone encoder.

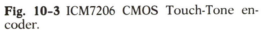
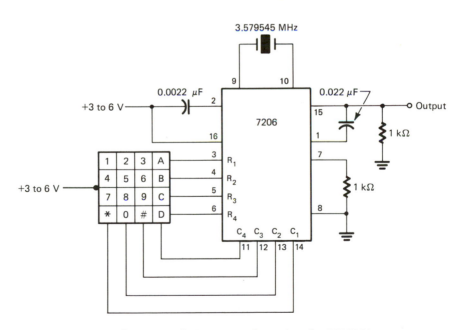

Fig. 10-4 Circuit for a Touch-Tone encoder using the ICM7206.

3½ DIGIT A/D CONVERTERS

All digital panel meters, such as a digital voltmeter, use some form of analog-to-digital (A/D) conversion. With the increasing use of computers for data acquisition, A/D converters are frequently used to process analog data. Among the popular CMOS devices available are Motorola's MC14433 and Intersil's ICL7106 (LCD drive) and ICL7107 (LED drive).

The MC14433, as shown by the pin diagram of Figure 10–5, is a single chip, 3½ digit A/D converter using a *modified dual ramp* technique of A/D conversion with a high input impedance, autopolarity, and autozero.

The digital output is a multiplexed 3½ digit BCD code with the most significant digit containing over- and underrange and polarity information, as well as the *half-digit* (blank, or 1). The MC14433 can be

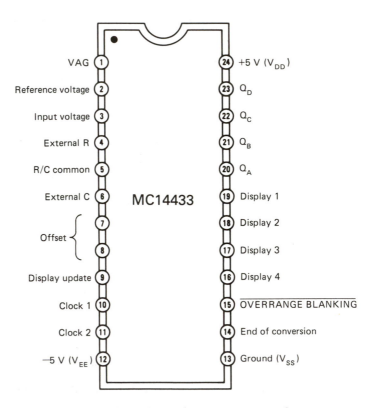

Fig. 10–5 MC14433 CMOS A/D converter pin diagram.

Table 10-3
MC14433 Reference Voltage and Integrator Resistors for Full-Scale Voltage

Full-Scale Voltage	Reference Voltage (pin 4)	Integrator Resistor (pins 4 & 5)
±1.999 mV	+200.0 mV	27 kΩ
±1.999 V	+2.000 V	470 kΩ

externally set to display a full-scale voltage of either ±199.9 mV or ±1.999 V by the values listed in Table 10–3 for the reference voltage and integrator resistor. A pair of 0.1-μF capacitors is used for the integrator (pins 5 and 6) and for offset correction (pins 7 and 8). Although an external clock may be used at pin 10, the MC14433 nevertheless contains its own internal clock (pin 11) so that a single resistor can be used between pins 10 and 11. This sets the frequency of the conversion cycle, which is composed of the unknown input measurement and auto-zero cycles.

The *end of conversion signal* (pin 14) indicates the end of one conversion cycle and the start of the next. In addition, the *display update signal* at pin 9 permits data to be strobed to the output latches. In most cases these two pins are tied together, thus allowing the output display to be updated at the end of each conversion cycle.

Figure 10–6 shows how the MC14433 is used to create a simple digital voltmeter with a full-scale reading of ±1.999 V. A 4511 device decodes the 4-bit BCD output to the 3½ digit *common-cathode* display. At pin 15, the *overrange blanking signal*, which is normally at logic 1, *blanks* the display when the input voltage exceeds ±2.000 V. A 75492 inverter/ driver is used for interfacing the four *display enable* outputs to the display during the strobing or multiplexing process. This circuit can then be enhanced by using the resistor divider string of Figure 10–7; the basic circuit is then able to display full-scale voltages from ±1.999 to ±1999.99 V.

Intersil's ICL7106 and 7107, whose pin diagrams are shown in Figure 10–8, are both 3½ digit A/D converters similar in function to the MC14433. The 7106 is designed to be interfaced with a *LCD* display and therefore includes the necessary backplane drive signal (pin 21). On the other hand, the 7107 will directly drive 0.3- or 0.43-in. common-anode *LED* displays.

The general circuits of Figures 10–9 (ICL7106) and 10–10 (ICL7107) are specified for a full-scale range of 200 mV. Like the

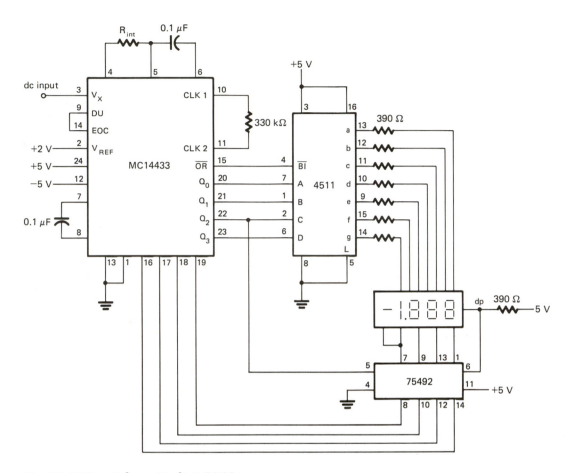

Fig. 10-6 Circuit for a 3½ digit DVM.

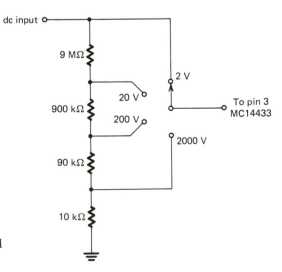

Fig. 10-7 Resistor divider for increasing DVM
full-scale voltage range.

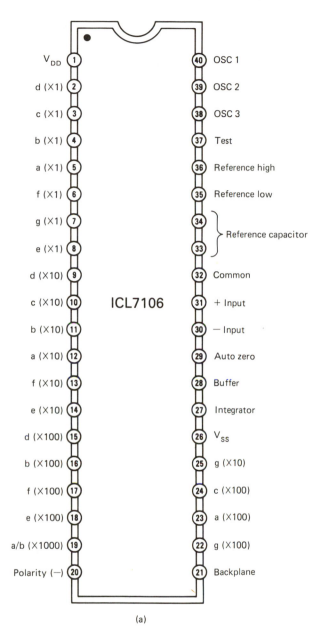

(a)

Fig. 10-8 Intersil CMOS A/D converters: (a) ICL7106 for LCD displays, (b) ICL7107 for LED displays.

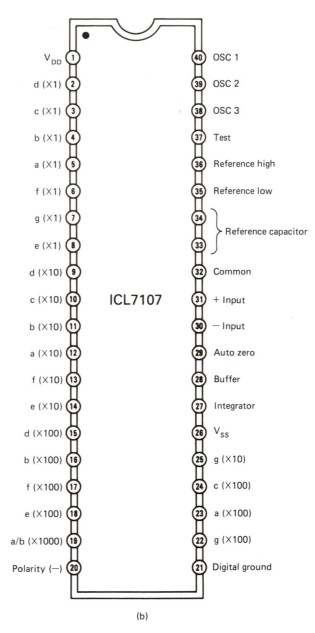

(b)

Fig. 10-8 (*Continued*)

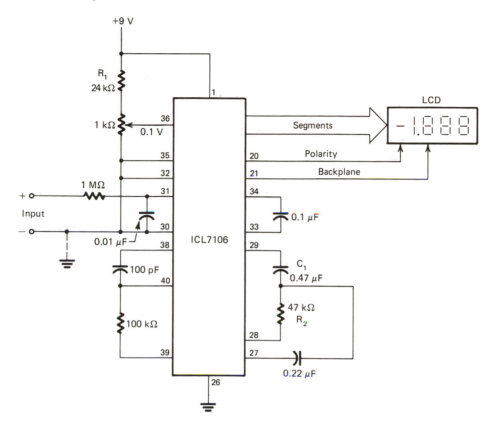

Fig. 10-9 Circuit for 3½ digit DVM with LCD display.

MC14433, either the 7106 or 7107 may be wired to permit a 2.000-V full-scale range by changing the values of the reference resistor R_1, the integrator resistor R_2, and the auto-zero capacitor C_1 to those values listed in Table 10–4.

The MC14433 and 7106/7107 devices are designed to operate from ±5-V supplies. However, when a negative supply is not

Table 10-4
ICL7106/7107 Required Component Values

Component	200 mV Full Scale	2.000 V Full Scale
R_1	24 kΩ	1.5 kΩ
R_2	47 kΩ	470 kΩ
C	0.47 μF	0.047 μF

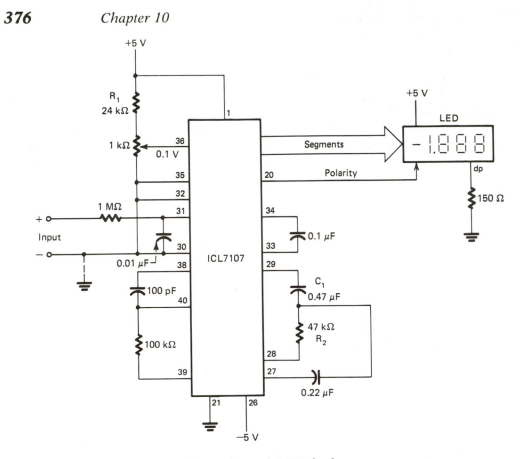

Fig. 10-10 Circuit for 3½ digit DVM with LED display.

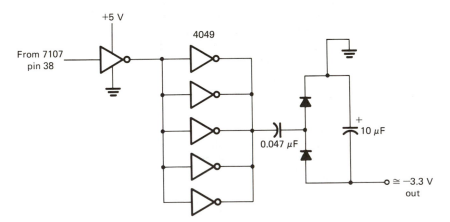

Fig. 10-11 Circuit for generating a negative supply voltage from a positive supply.

convenient, as in certain portable applications, a suitable substitute is shown in Figure 10–11.

4046 CMOS PHASE-LOCKED LOOP

The 4046 phase-locked loop[*] whose pin diagram is shown in Figure 10–12 and block diagram in Figure 10–13, contains two phase detectors and a voltage-controlled oscillator (VCO). All that is needed is an external low-pass loop filter.

The two phase detectors have their inputs connected in parallel, although the outputs are brought out separately. Phase detector I, sometimes referred to as the *low-noise detector*, is basically an XOR gate. As shown by the timing diagram of an XOR-gate phase detector in Figure 10–14, *both the input and VCO signals must be 50% duty cycle (i.e., symmetrical) square waves.*

Phase detector II, often called the *wideband detector*, is an edge-triggered type, which triggers on the positive leading edges of its two

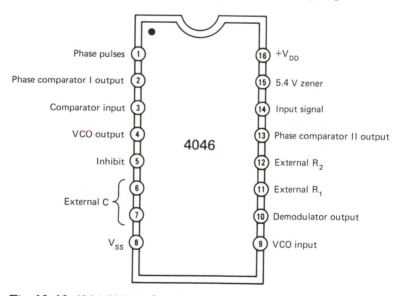

Fig. 10–12 4046 CMOS phase-locked loop pin diagram.

---

[*]Additional details about the 4046 and other aspects of the phase-locked loop are found in H. M. Berlin, *Phase-Locked Loop Circuits, with Experiments,* Howard W. Sams & Co., Inc., Indianapolis, 1978. A rigorous mathematical treatment can be found in F. M. Gardner, *Phaselock Techniques,* 2nd ed., John Wiley & Sons, Inc., New York, 1979.

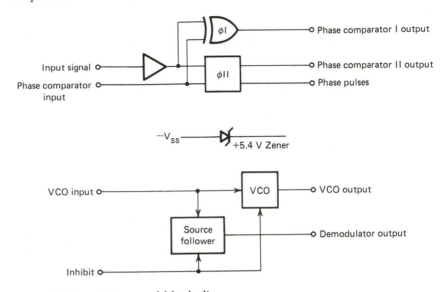

Fig. 10-13 4046 internal block diagram.

inputs. If the input signal, which can be a pulse train having any duty cycle, is *lower* than the VCO frequency, the output is at logic 0. On the other hand, if the input frequency is higher than the VCO, the output is at logic 1. If both frequencies are equal, the output of the phase detector is a pulse whose width is proportional to the phase difference between the input and VCO. As shown in Figure 10–14, this pulse is positive when the VCO *lags* the input (or input leads the VCO) and negative when the VCO leads the input. One advantage of the edge-triggered phase detector over the XOR type is that the edge-triggered

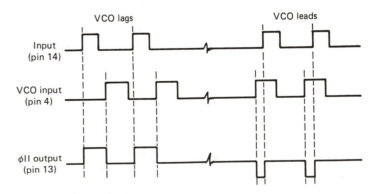

Fig. 10-14 4046 phase comparator II timing diagram.

detector is insensitive to multiples (harmonics) of the input frequency, while the XOR type may tend to lock onto these harmonic frequencies.

Also shown in the block diagram is an uncommitted 5.4-V zener diode, which can be used if regulation of the power supply voltage is necessary. The VCO, sometimes called a voltage-to-frequency (V/F) converter, is an oscillator whose output frequency is proportional to its input voltage. Therefore, the VCO output frequency is a minimum value when the input voltage is zero and increases *linearly* to a maximum value when the input equals the supply voltage. As shown in the basic circuit of Figure 10–15, the minimum–maximum VCO range is set by R_1 and C_1, while the minimum VCO frequency is controlled by R_2 and C_1. Momentarily grounding pin 9 causes the VCO to oscillate at its minimum frequency, while connecting pin 9 to the positive supply voltage produces the maximum frequency. If, for certain applications, it is not necessary to have the VCO on at certain times, the VCO can be turned off by tying the *inhibit* input (pin 5) to logic 1, which will in turn also minimize power consumption. Otherwise, the *inhibit* input is normally connected to ground.

To get the 4046 or any other phase-locked loop system to work, a *low-pass loop filter* is required between the output of the phase detector and the VCO input. Although the loop filter can take on several forms, the *lead–lag* circuit of Figure 10–16 is frequently used. The ratio of R_1/R_2 basically determines the damping of the loop, while the $R_1 C$

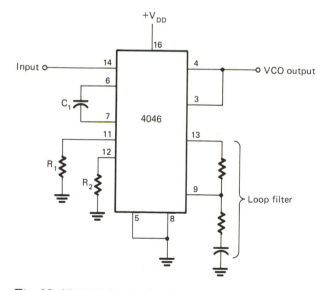

Fig. 10–15 4046 basic circuit.

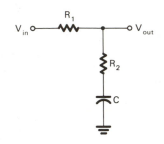

Fig. 10-16 Lag–lead low-pass loop filter.

time constant determines the *settling time.* Generally, the value of R_1 is made anywhere from three to ten times larger than R_2.

The *phase pulses* output (pin 1) of phase detector II can be NORed with the output of phase detector (pin 2) to form a lock detector, as shown in Figure 10–17. When the loop is locked, the output of the last NOR gate will be at logic 1.

By breaking the feedback path between the VCO output and one input of the phase detector (pins 4 and 3) and inserting a divide-by-N counter, the 4046 can be used as an effective frequency synthesizer or multiplier. As an example, the resolution of a typical frequency counter can be improved from 1 Hz to 0.01 Hz for low-frequency signals by using a pair of decade counters (4017, 74C90, etc.) in cascade, as shown

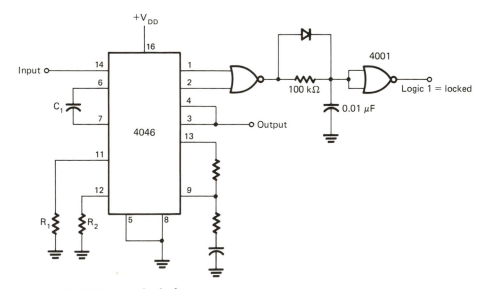

Fig. 10-17 NOR gate lock detector.

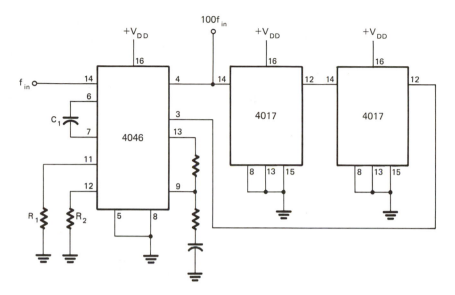

Fig. 10-18 Phase-locked loop frequency × 100 synthesizer or prescaler.

in Figure 10–18. Thus an input frequency of 58.23 Hz will be displayed as 5823 Hz on the counter. Otherwise, all you would read on the counter's display is either 58 or 59 Hz. By making the modulus of the counter chain equal to 60, this same technique is then used to display the input frequency in terms of *cycles per minute*. This is extremely useful when measuring the frequency of physiological events such as respiration (breaths per minute) and heart rate (beats per minute).

TWO-CHIP FREQUENCY COUNTER

Chapter 7 discussed a method of combining a series of TTL decade counters, latches, decoder drivers, and LED displays to form a seven-digit frequency counter for input frequencies up to 10 MHz with a 1-Hz resolution. In all, 45 individual elements were required, which did not include the crystal-controlled time base. Such a design, although a straightforward and traditional approach, nevertheless would physically occupy too much space and require at least 0.5 A at 5 V. Consequently, such a design would not be practical for portable operation.

Using both the Intersil ICM7207 oscillator controller and the ICM7208 seven-decade counter decoder/driver, we are able to build a smaller and more efficient battery-powered counter capable of frequency measurements. Shown in Figure 10–19a, the ICM7207 is a crystal-controlled oscillator having outputs for the count, latch, reset, and display multiplex functions required by the 7208. When a 6.5536-MHz crystal is used, it is divided by 2^{12} to produce the 1600-Hz signal that multiplexes the counter's display. The 1600-Hz signal is divided

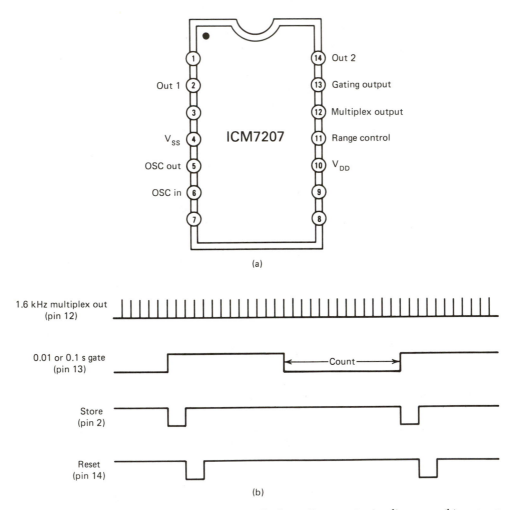

(a)

Fig. 10-19 ICM7207 CMOS crystal-controlled oscillator: (a) pin diagram, (b) output wave forms.

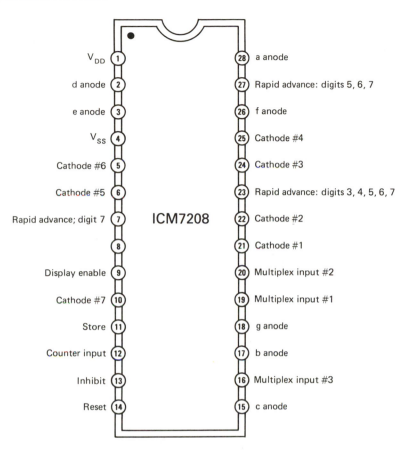

Fig. 10-20 ICM7208 CMOS 7-decade counter pin diagram.

further to provide counting gate signals of either 0.01 or 0.1 second. The timing diagram for the four output signals is shown in Figure 10–19b.

The ICM7208, shown in Figure 10–20, provides the following functions:

- A seven-decade counter and decoder/driver
- Multiplexer
- Logic for display reset, blanking, input inhibit, and display on/off

The complete two-chip counter is shown in Figure 10–21. Although the upper range of the counter is typically 2 MHz with a resolution of 1 Hz, this range can be increased to 20 or 200 MHz by the addition of a

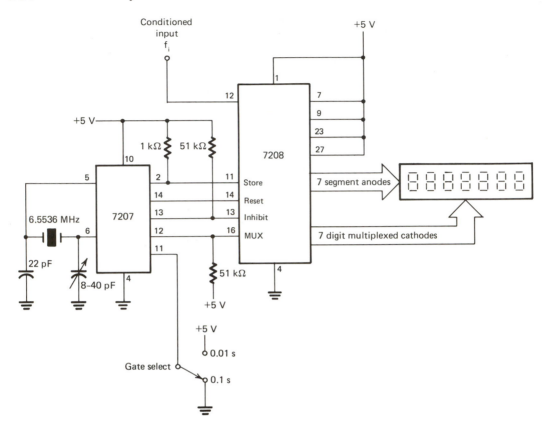

Fig. 10-21 Two-chip CMOS frequency counter.

suitable divide-by-10 or a divide-by-100 counter at the input (pin 12, ICM7208). Consequently, the resolution will be 10 Hz and 100 Hz, respectively. On the other hand, the resolution of the counter circuit can be increased at low frequencies with the addition of a 4046 phase-locked loop multiplier, as described in the previous section.

INTRODUCTION TO THE EXPERIMENTS

As an example of the use and operation of the specialized CMOS devices in this chapter, the following experiments use the 4046 phase-locked loop as representative. The experiments that you will perform can be summarized as follows:

Experiment	Purpose
10–1	Demonstrates the operation of the 4046 CMOS phase-locked loop.
10–2	Demonstrates the operation of a loss-of-lock indicator with the 4046 phase-locked loop.
10–3	Demonstrates the operation of a simple frequency synthesizer that can be used as a multiplying prescaler for frequency counters.

EXPERIMENT 10–1

PURPOSE

The purpose of this experiment is to demonstrate the operation of the 4046 CMOS phase-locked loop.

PIN CONFIGURATION OF INTEGRATED-CIRCUIT DEVICE

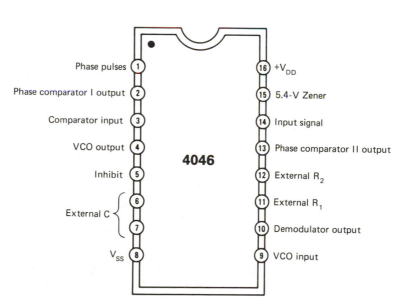

Fig. 10–22 4046 pin diagram.

Required Parts and Equipment

1   560-Ω resistor
1   4.7-kΩ resistor
1   27-kΩ resistor
1   100-kΩ resistor
2   0.1-μF capacitor
1   4046 CMOS phase-locked loop
1   solderless breadboard
1   power supply (5 V)
1   function generator
1   oscilloscope
1   frequency counter

SCHEMATIC DIAGRAM OF CIRCUIT

STEP 1

Set your oscilloscope to the following settings:

- Channel 1: 0.5 V/division
- Time base: 0.5 ms/division

STEP 2

Wire the circuit shown in the schematic diagram of Figure 10–23 and apply power to the breadboard. Adjust the output of the function generator (sine wave) at approximately 1 kHz with a frequency counter and the peak-to-peak voltage to 2 V (i.e., four vertical divisions). Then connect the frequency counter to pins 3 and 4 of the 4046 device. What do you notice about the output frequency of the phase-locked loop?

The output frequency of the phase-locked loop should be the same as its sine-wave input.

STEP 3

With an extra piece of wire, connect pin 9 of the 4046 to *ground*. Record the resultant output frequency of the phase-locked loop:

$$f_L = \text{_____} \text{ Hz}$$

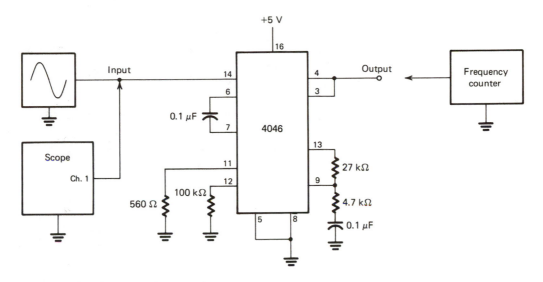

Fig. 10-23 Schematic diagram for Experiment 10–1.

This output frequency is the lower limit of the VCO, which is determined by the 0.1-μF capacitor connected between pins 6 and 7 and the 100-kΩ resistor connected between pin 12 and ground.

STEP 4

Now, with the same wire, connect pin 9 to the +5-V supply. Record the resultant output frequency:

$$f_U = \rule{2cm}{0.4pt} \text{ Hz}$$

This output should be higher than the one measured in step 3. This is the upper limit of the VCO, which is basically determined by the 560-Ω resistor and the 0.1-μF capacitor between pins 6 and 7.

STEP 5

Now remove this connection between pin 9 and the +5-V supply. You should again measure an output frequency that is the same as the frequency of the function generator as in step 2.

STEP 6

Connect the frequency counter to pins 3 and 4 of the 4046. Now slowly increase the frequency of the function generator. What do you observe on the frequency counter?

You should observe that the output frequency also increases. In fact, the output frequency follows the changes in the input frequency and should be exactly equal. Measure the input to confirm this.

STEP 7

While watching the output frequency of the phase-locked loop, continue to slowly increase the input frequency, stopping when the output frequency does not increase any further. Measure the input frequency and record your result:

$$f_{IN}(H) = \underline{\hspace{2cm}} \text{ Hz}$$

You should find that this frequency is about the same as the frequency you measured in step 4, the *upper frequency limit of the VCO*. The phase-locked loop then follows input frequency changes for frequencies below this upper limit.

STEP 8

Now decrease the input frequency while observing the frequency counter. At some point the output frequency will remain constant although the input frequency is being manually decreased. Measure this output frequency and record your result:

$$f_{IN}(L) = \underline{\hspace{2cm}} \text{ Hz}$$

You should find that this frequency is about the same as the frequency that you measured in step 3, the *lower frequency limit of the VCO*. Consequently, the phase-locked loop follows changes in the input frequency for any frequency between the lower and upper frequency limits of the VCO.

In this range, the loop is said to be *locked,* and the range over which the phase-locked loop follows changes in the output frequency is called the *lock range.* To determine the lock range from your measurements, subtract the value measured in step 8 from the value of step 7, and record your result:

Lock range = _____ Hz

This lock range can be easily changed by simply changing the value of the resistor connected to pin 11 or 12. Decreasing the 100-kΩ resistor at pin 12, for example, *increases the lower frequency*. Keep this circuit on your breadboard, as it will be used in the next experiment.

EXPERIMENT 10–2

PURPOSE

The purpose of this experiment is to demonstrate a *loss-of-lock* indicator with the phase-locked loop circuit of Experiment 10–1.

**PIN CONFIGURATION OF
INTEGRATED-CIRCUIT DEVICES**

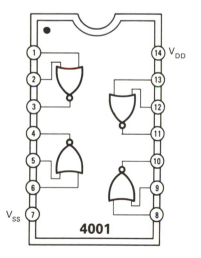

Fig. 10–24 4001 pin diagram.

Required Parts and Equipment

Those required for Experiment 10–1 in addition to:

1   100-kΩ resistor
1   0.01-μF capacitor
1   1N914 diode
1   4001 CMOS quad NOR gate
1   LED monitor

SCHEMATIC DIAGRAM OF CIRCUIT

STEP 1

Wire the loss-of-lock circuit as shown in the schematic diagram. Connect pin 1 of the 4001 CMOS NOR gate to pin 1 of the 4046 phase-locked loop (*phase pulses* output of comparator II) and pin 2 of the 4001 NOR gate to pin 2 of the 4046 (phase comparator I). Make sure that you have connected the 1N914 diode correctly across the 100-kΩ resistor.

STEP 2

Apply power to the breadboard and set the input frequency at approximately 500 Hz. Is the LED monitor lit or unlit?

The LED monitor should be *lit* since the 500-Hz input frequency is within the lock range of the loop circuit. This was

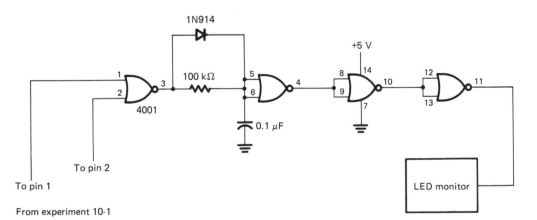

Fig. 10–25 Schematic diagram for Experiment 10–2.

determined in steps 7 and 8 of the previous experiment. When the loop is phase locked, the output of the loss-of-lock circuit is at logic 1.

STEP 3

Increase the input frequency just past the VCO's upper limit (Experiment 10–1, step 7). What happens to the LED monitor?

The LED monitor should now be *unlit*, indicating that the phase-locked loop is now *unlocked*, as the input frequency is now outside the lock range of the loop. In some cases, we have noticed that the LED monitor flashes several times as the loop becomes unlocked. This is due to the transient behavior of the loop.

STEP 4

Change the input frequency to 1 kHz. From the LED monitor, is the loop locked or unlocked?

The LED monitor should be *lit* since the 1-kHz input signal is within the lock range of the loop. We can use such a circuit with the 4046 phase-locked loop to visually indicate whether the loop is locked or not.

EXPERIMENT 10–3

PURPOSE

The purpose of this experiment is to demonstrate the operation of a simple frequency synthesizer that can be used as a multiplying prescaler for frequency counters.

PIN CONFIGURATION OF
INTEGRATED-CIRCUIT DEVICES

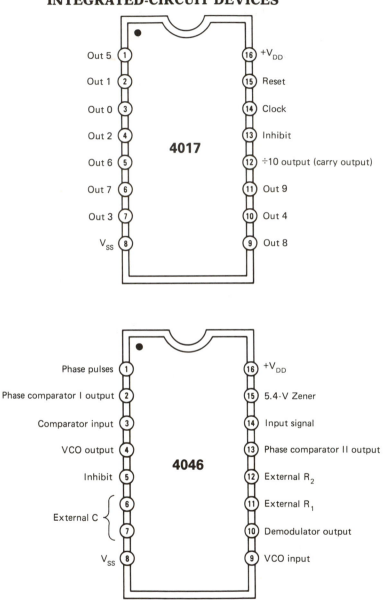

Fig. 10-26 4017 and 4046 pin diagrams.

Required Parts and Equipment

1   560-Ω resistor

1   4.7-kΩ resistor

1   27-kΩ resistor

1   100-kΩ resistor

2   0.1-μF capacitors

1   4017 CMOS counter

1   4046 CMOS phase-locked loop

1   solderless breadboard

1   power supply (5 V)

1   function generator

1   oscilloscope

1   frequency counter

SCHEMATIC DIAGRAM OF CIRCUIT

STEP 1

Set your oscilloscope for the following settings:

- Channel 1: 0.5 V/division
- Time base: 10 ms/division

STEP 2

Wire the circuit shown in the schematic diagram of Figure 10–27 and apply power to the breadboard. Connect the frequency counter to pin 14 of the 4046 phase-locked loop. Adjust the input frequency so that it is somewhere between 80 and 90 Hz. In addition, adjust the peak-to-peak input sine wave at 2 V.

STEP 3

Measure the input frequency and record your result:

$$f_i = \underline{\hspace{2cm}} \text{ Hz}$$

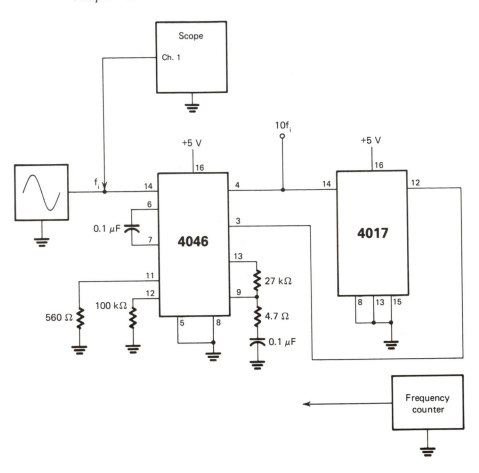

Fig. 10-27 Schematic diagram for Experiment 10–3.

STEP 4

Now connect the frequency counter to pin 4 of the phase-locked loop. Measure the *output* and record your result:

$$f_o = \underline{\hspace{3cm}} \text{ Hz}$$

What relationship do you notice about the input and output frequencies of the phase-locked loop?

You should have observed that the output frequency is exactly ten times larger than the input frequency. Most frequency counters for this input frequency range normally have a resolution of ± 1 Hz. By using this circuit to multiply the frequency counter's input by 10, you are then able to measure frequencies with a resolution of ± 0.1 Hz. As an example, if you measured an input frequency of 87 Hz in step 3, the actual frequency could be anywhere from 86 to 88 Hz. If the measured output frequency of the phase-locked loop in step 4 was 867 Hz, then the input frequency is precisely *86.7* Hz, not 87 Hz as was probably first measured in step 3. The resolution of your frequency counter has now been increased by one significant digit.

STEP 5

Now adjust the input frequency to any value between 200 and 300 Hz, measuring both the input and output frequencies of the phase-locked loop (pins 14 and 4, respectively). Assuming that you are still within the lock range of the phase-locked loop, you should find that, whatever the input frequency is, the output is always ten times larger than the input. If not, the loop is now unlocked and the output frequency is the free-running frequency of the VCO.

APPENDIX A: TTL–CMOS FUNCTIONAL EQUIVALENTS

TTL	CMOS	TTL	CMOS
7400	4011, 74C00	74107	4027, 74C107
7401	40107	74110	4095
7402	4001, 74C02	74111	4027
7404	4009, 4049, 74C04	74114	40106
7406	4009, 4049	74121	4047, 4098
7407	4010, 4050	74122	4047, 4098
7408	4081, 74C08	74123	4098, 74C221
7410	4023, 74C10	74125	4502
7414	MC14584, 40106, 74C14	74126	4502
7420	4012, 74C20	74132	4093
7425	4002	74136	4030, 4070
7427	4025	74141	4028
7428	4001	74145	4028
7430	4068, 74C30	74148	4532
7432	4071, 74C32	74150	4067
7437	4011	74151	4051, 4097, 74C151
7440	4012	74152	4051, 4097
7442	4028, 74C42	74153	4052
7445	4028	74154	4514, 4515, 74C154
7446	4055, 4511	74155	4555, 4556
7447	4055, 4511	74156	4555, 4556
7448	4511, 74C48	74157	4019, 74C157

(Continued on next page)

TTL	CMOS	TTL	CMOS
7449	4511	74160	74C160
7450	4085	74161	74C161
7453	4086	74162	74C162
7454	4086	74163	74C163
7470	4096	74164	4015, 74C164
7472	4095	74165	4021, 74C165
7473	4027, 74C73	74166	4014
7474	4013, 74C74	74167	4527
7475	4042	74173	4076, 74C173
7476	4027, 74C76	74174	74C174
7477	4042	74175	74C175
7478	44027	74178	4035
7483	4008, 74C83	74179	4035
7485	4063, 74C85	74180	40101
7486	4030, 4070, 74C86	74181	40181
7489	74C89	74182	40182
7490	4518, 74C90	74190	4510
7491	4015, 4094	74191	4516
7493	4520, 74C93	74192	40192, 74C192
7494	4035	74193	40193, 74C193
7495	40104, 40194, 74C95	74194	40104, 40194
7499	40104, 40194	74195	4035, 40195, 74C195
74100	4034	74198	4034
74104	4095	74200	4061, 74C200
74105	4095		

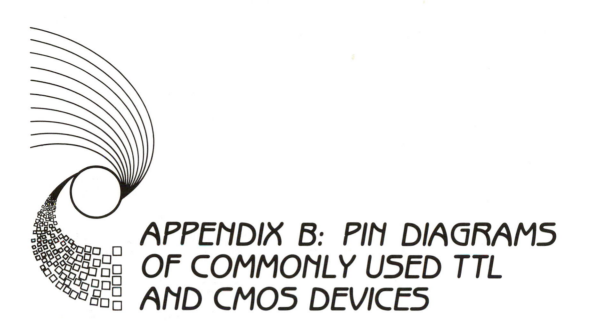

APPENDIX B: PIN DIAGRAMS OF COMMONLY USED TTL AND CMOS DEVICES

The following pages show, as a handy reference, pin diagrams of the most commonly used TTL and CMOS devices. 74C series CMOS devices are not shown as they are pin-for-pin equivalents of the 7400 series TTL devices.

TTL DEVICES

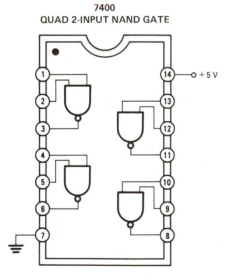

7400
QUAD 2-INPUT NAND GATE

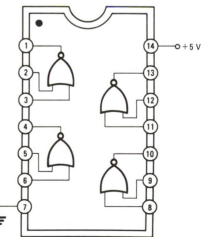

7402
QUAD 2-INPUT NOR GATE

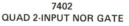

7404
HEX INVERTER

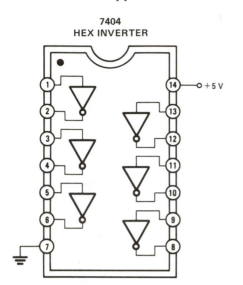

7408
QUAD 2-INPUT AND GATE

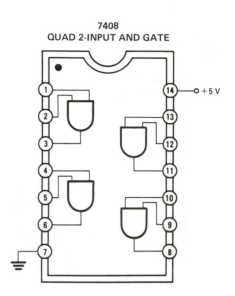

7410
TRIPLE 3-INPUT NAND GATE

7414
HEX SCHMITT TRIGGER
(Inverting)

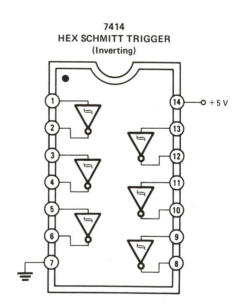

7420
DUAL 4-INPUT NAND GATE

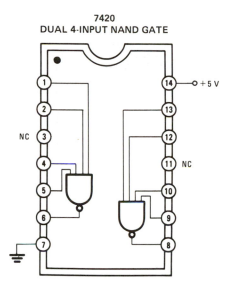

7430
8-INPUT NAND GATE

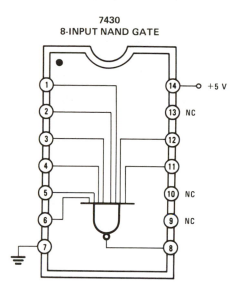

7432
QUAD 2-INPUT OR GATE

7442
BCD TO 1-OF-10 DECODER

7445
BCD TO 1-OF-10
DECODER/DRIVER

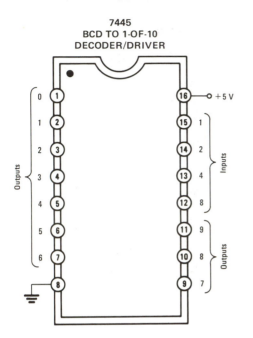

7447
BCD TO 7-SEGMENT
DECODER-DRIVER

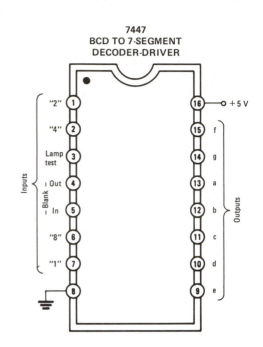

7473
DUAL JK
LEVEL-TRIGGERED FLIP-FLOP
With Preclear Only

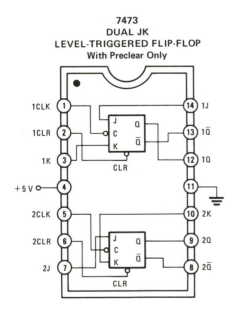

7474
DUAL D
EDGE-TRIGGERED FLIP-FLOP
With Preset and Preclear

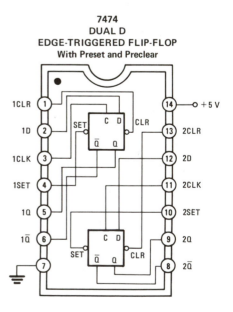

7475
QUAD LATCH

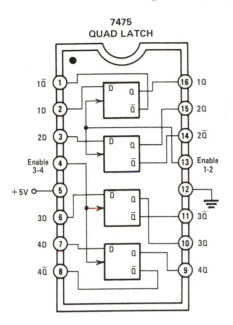

7476
DUAL JK
LEVEL-TRIGGERED FLIP-FLOP
With Preset and Preclear

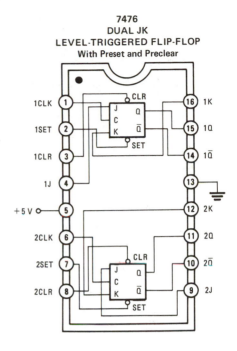

7483
4-BIT FULL ADDER

7485
4-BIT MAGNITUDE COMPARATOR

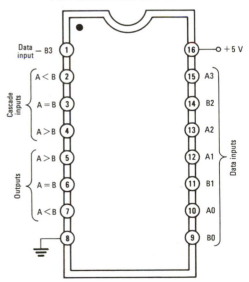

7486
QUAD EXCLUSIVE-OR GATE

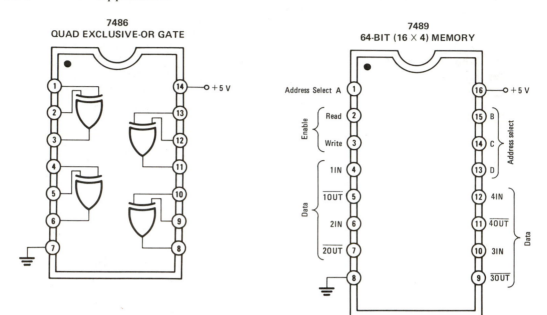

7489
64-BIT (16 × 4) MEMORY

7490
DECADE COUNTER (÷10)

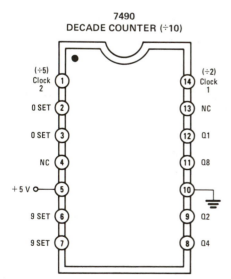

7492
BASE-TWELVE (÷12) COUNTER

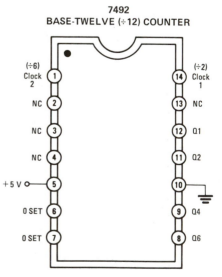

7493
BINARY (÷16) COUNTER

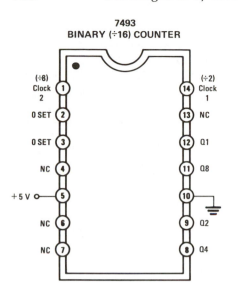

7495
SHIFT REGISTER
4 BITS Right-Left
Parallel In, Parallel Out

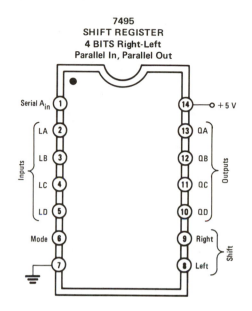

7496
SHIFT REGISTER,
5 BITS Shift Right,
Parallel In, Parallel Out

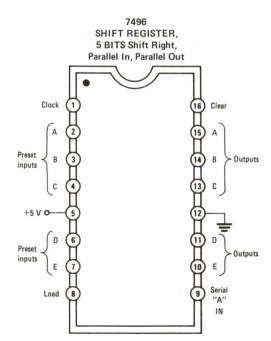

74107
DUAL JK
LEVEL-TRIGGERED FLIP-FLOP
With Preclear Only

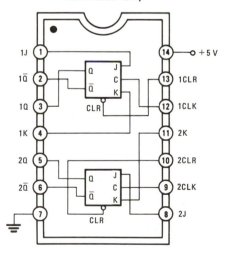

74121
MONOSTABLE MULTIVIBRATOR
Single, Not Retriggerable

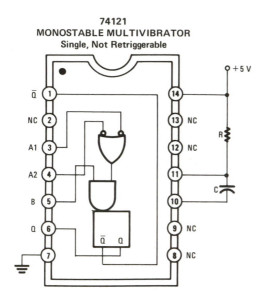

74122
MONOSTABLE MULTIVIBRATOR
Single, Retriggerable

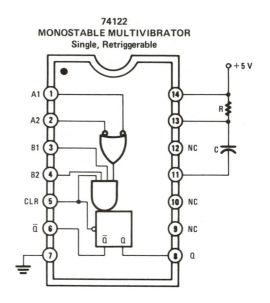

74123
MONOSTABLE MULTIVIBRATOR
Dual, Retriggerable

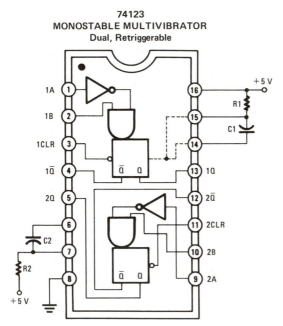

74125
QUAD TRI-STATE DRIVER
(Low Enable)

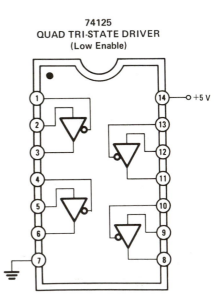

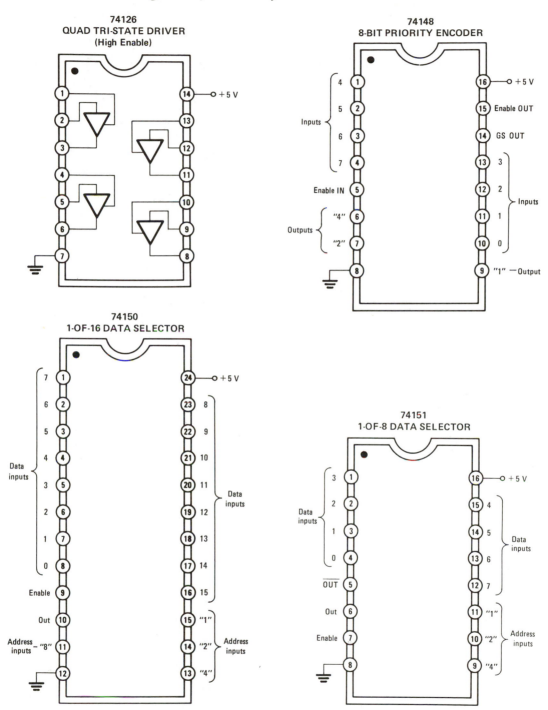

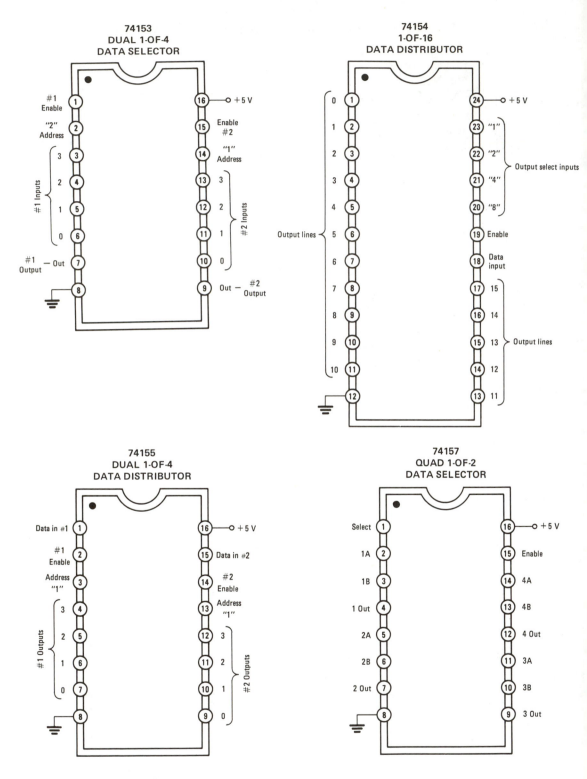

74153
DUAL 1-OF-4
DATA SELECTOR

#1 Enable	1	16 — +5 V
"2" Address	2	15 — Enable #2
3	3	14 — "1" Address
2	4	13 — 3
1	5	12 — 2
0	6	11 — 1
#1 — Out Output	7	10 — 0
	8	9 — Out — #2 Output

#1 Inputs (pins 3–6)
#2 Inputs (pins 13–10)

74154
1-OF-16
DATA DISTRIBUTOR

0	1	24 — +5 V
1	2	23 — "1"
2	3	22 — "2"
3	4	21 — "4"
4	5	20 — "8"
5	6	19 — Enable
6	7	18 — Data input
7	8	17 — 15
8	9	16 — 14
9	10	15 — 13
10	11	14 — 12
	12	13 — 11

Output lines (0–10)
Output select inputs ("1", "2", "4", "8")
Output lines (15–11)

74155
DUAL 1-OF-4
DATA DISTRIBUTOR

Data in #1	1	16 — +5 V
#1 Enable	2	15 — Data in #2
Address "1"	3	14 — #2 Enable
3	4	13 — Address "1"
2	5	12 — 3
1	6	11 — 2
0	7	10 — 1
	8	9 — 0

#1 Outputs (pins 4–7)
#2 Outputs (pins 12–9)

74157
QUAD 1-OF-2
DATA SELECTOR

Select	1	16 — +5 V
1A	2	15 — Enable
1B	3	14 — 4A
1 Out	4	13 — 4B
2A	5	12 — 4 Out
2B	6	11 — 3A
2 Out	7	10 — 3B
	8	9 — 3 Out

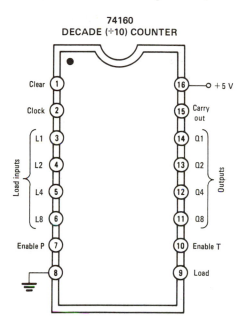

74160
DECADE (÷10) COUNTER

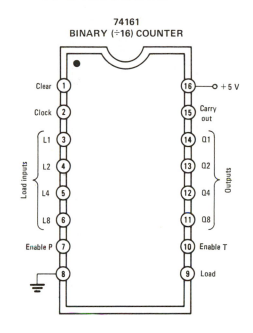

74161
BINARY (÷16) COUNTER

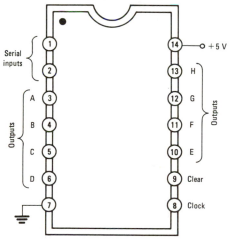

74164
SHIFT REGISTER, 8 BITS
Serial Input, Parallel Output

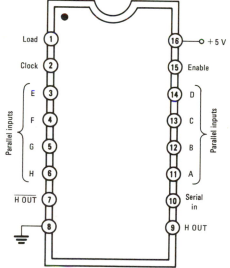

74165
SHIFT REGISTER, 8 BITS
Parallel Input, Serial Output

74175
QUAD "D" LATCH

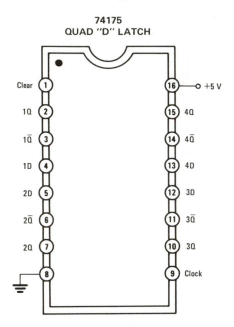

Pin	Signal
1	Clear
2	1Q
3	1$\overline{\text{Q}}$
4	1D
5	2D
6	2$\overline{\text{Q}}$
7	2Q
8	(ground)
16	+5 V
15	4Q
14	4$\overline{\text{Q}}$
13	4D
12	3D
11	3$\overline{\text{Q}}$
10	3Q
9	Clock

74176
DECADE (÷10) COUNTER

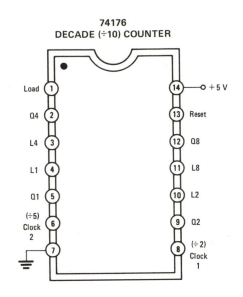

Pin	Signal
1	Load
2	Q4
3	L4
4	L1
5	Q1
6	(÷5) Clock 2
7	(ground)
14	+5 V
13	Reset
12	Q8
11	L8
10	L2
9	Q2
8	(÷2) Clock 1

74190
DECADE (÷10)
UP/DOWN COUNTER

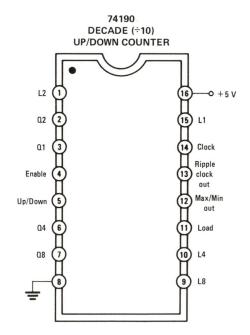

Pin	Signal
1	L2
2	Q2
3	Q1
4	Enable
5	Up/Down
6	Q4
7	Q8
8	(ground)
16	+5 V
15	L1
14	Clock
13	Ripple clock out
12	Max/Min out
11	Load
10	L4
9	L8

74191
BINARY (÷16)
UP/DOWN COUNTER

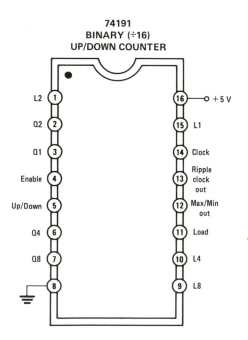

Pin	Signal
1	L2
2	Q2
3	Q1
4	Enable
5	Up/Down
6	Q4
7	Q8
8	(ground)
16	+5 V
15	L1
14	Clock
13	Ripple clock out
12	Max/Min out
11	Load
10	L4
9	L8

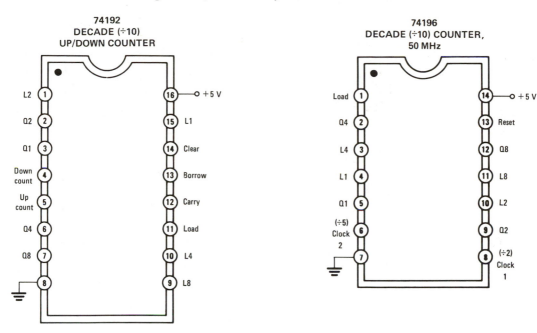

CMOS DEVICES

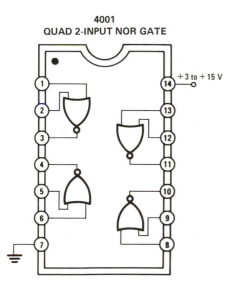

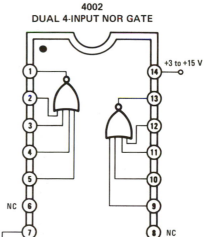

4008
4-BIT FULL ADDER

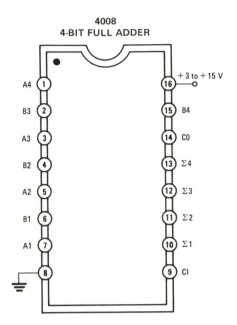

4011
QUAD 2-INPUT NAND GATE

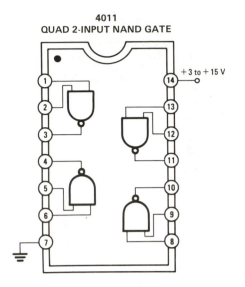

4012
DUAL 4-INPUT NAND GATE

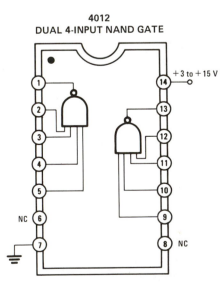

4013
DUAL D FLIP-FLOP

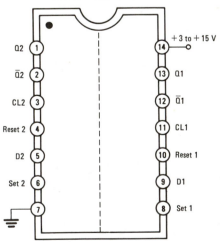

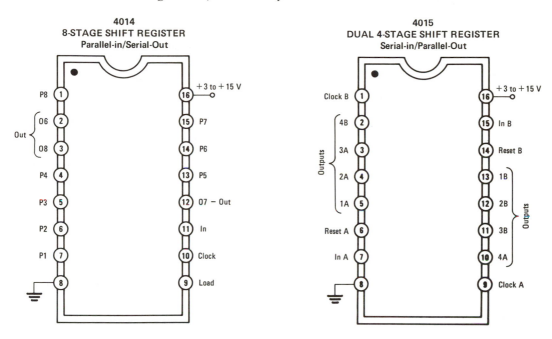

4014
8-STAGE SHIFT REGISTER
Parallel-in/Serial-Out

Pin	Signal
P8 (1)	(16) +3 to +15 V
O6 (2)	(15) P7
O8 (3)	(14) P6
P4 (4)	(13) P5
P3 (5)	(12) O7 — Out
P2 (6)	(11) In
P1 (7)	(10) Clock
(8)	(9) Load

Out { O6, O8

4015
DUAL 4-STAGE SHIFT REGISTER
Serial-in/Parallel-Out

Pin	Signal
Clock B (1)	(16) +3 to +15 V
4B (2)	(15) In B
3A (3)	(14) Reset B
2A (4)	(13) 1B
1A (5)	(12) 2B
Reset A (6)	(11) 3B
In A (7)	(10) 4A
(8)	(9) Clock A

Outputs { 4B, 3A, 2A, 1A
Outputs } 1B, 2B, 3B, 4A

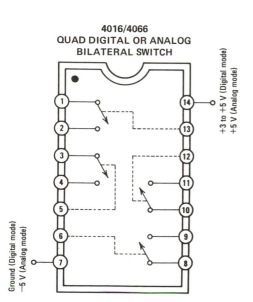

4016/4066
QUAD DIGITAL OR ANALOG
BILATERAL SWITCH

+3 to +5 V (Digital mode)
+5 V (Analog mode)

Ground (Digital mode)
−5 V (Analog mode)

4017
DIVIDE-BY-10 COUNTER WITH
1-OF-10 OUTPUTS

Pin	Signal
5 (1)	(16) +3 to +15 V
1 (2)	(15) Reset
0 (3)	(14) Clock
2 (4)	(13) Enable
6 (5)	(12) Out ÷10
7 (6)	(11) 9
3 (7)	(10) 4
(8)	(9) 8

Outputs { 5, 1, 0, 2, 6, 7, 3
Outputs } 9, 4, 8

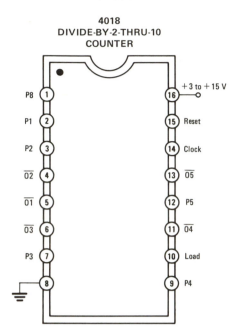

4018
DIVIDE-BY-2-THRU-10
COUNTER

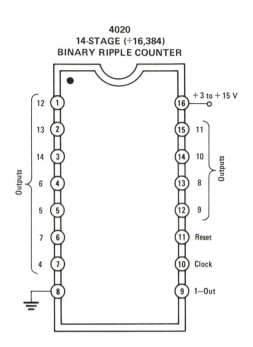

4020
14-STAGE (÷16,384)
BINARY RIPPLE COUNTER

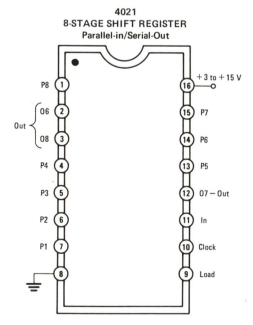

4021
8-STAGE SHIFT REGISTER
Parallel-in/Serial-Out

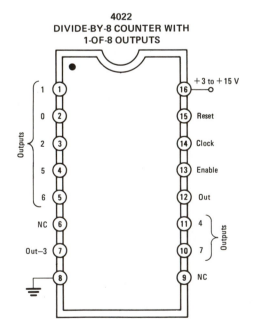

4022
DIVIDE-BY-8 COUNTER WITH
1-OF-8 OUTPUTS

4023
TRIPLE 3-INPUT NAND GATE

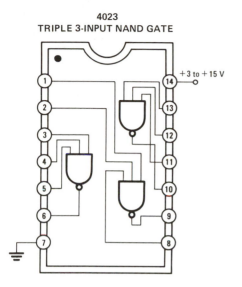

4024
7-STAGE (÷128) BINARY
RIPPLE COUNTER

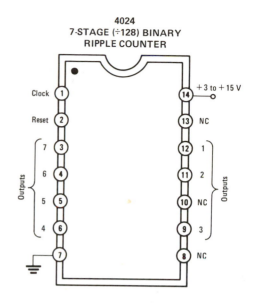

4025
TRIPLE 3-INPUT NOR GATE

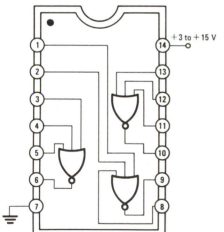

4026
DECADE (÷10) COUNTER WITH
7-SEGMENT DECODED OUTPUT

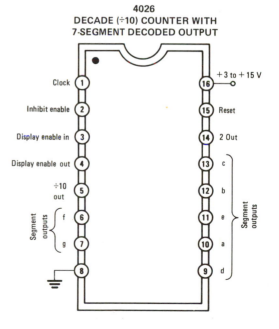

4027
DUAL JK FLIP-FLOP

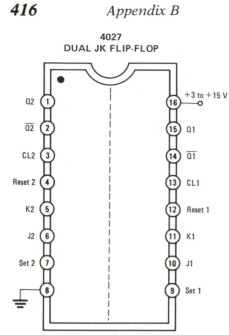

Q2 — 1
Q̄2 — 2
CL2 — 3
Reset 2 — 4
K2 — 5
J2 — 6
Set 2 — 7
8

16 — +3 to +15 V
15 — Q1
14 — Q̄1
13 — CL1
12 — Reset 1
11 — K1
10 — J1
9 — Set 1

4028
BCD TO DECIMAL
(1-of-10) DECODER

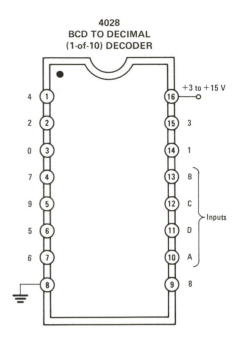

4 — 1
2 — 2
0 — 3
7 — 4
9 — 5
5 — 6
6 — 7
8

16 — +3 to +15 V
15 — 3
14 — 1
13 — B
12 — C ⎫ Inputs
11 — D
10 — A ⎭
9 — 8

4029
UP-DOWN DIVIDE-BY-10 OR
DIVIDE-BY-16 COUNTER

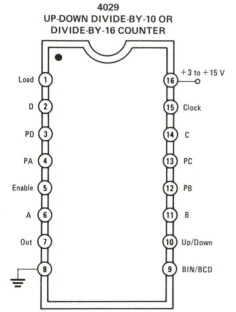

Load — 1
D — 2
PD — 3
PA — 4
Enable — 5
A — 6
Out — 7
8

16 — +3 to +15 V
15 — Clock
14 — C
13 — PC
12 — PB
11 — B
10 — Up/Down
9 — BIN/BCD

4030
QUAD EXCLUSIVE OR GATE

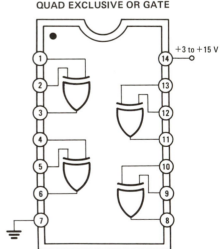

1
2
3
4
5
6
7

14 — +3 to +15 V
13
12
11
10
9
8

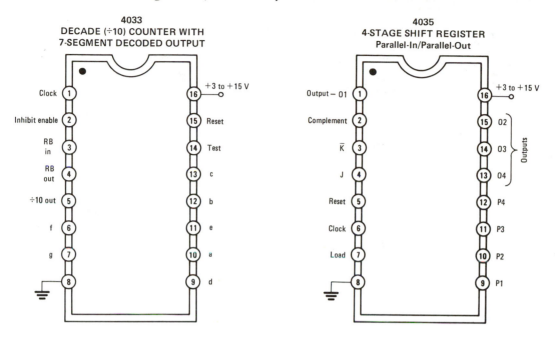

4033
DECADE (÷10) COUNTER WITH
7-SEGMENT DECODED OUTPUT

Clock	1		16	+3 to +15 V
Inhibit enable	2		15	Reset
RB in	3		14	Test
RB out	4		13	c
÷10 out	5		12	b
f	6		11	e
g	7		10	a
	8		9	d

4035
4-STAGE SHIFT REGISTER
Parallel-In/Parallel-Out

Output — Q1	1		16	+3 to +15 V
Complement	2		15	Q2
$\overline{K}$	3		14	Q3
J	4		13	Q4
Reset	5		12	P4
Clock	6		11	P3
Load	7		10	P2
	8		9	P1

Outputs (Q2, Q3, Q4)

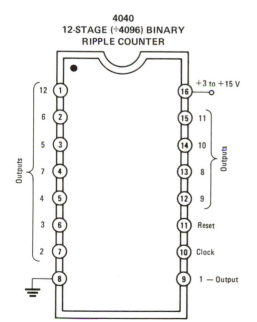

4040
12-STAGE (÷4096) BINARY
RIPPLE COUNTER

12	1		16	+3 to +15 V
6	2		15	11
5	3		14	10
7	4		13	8
4	5		12	9
3	6		11	Reset
2	7		10	Clock
	8		9	1 — Output

Outputs

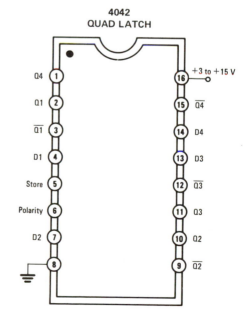

4042
QUAD LATCH

Q4	1		16	+3 to +15 V
Q1	2		15	$\overline{Q4}$
$\overline{Q1}$	3		14	D4
D1	4		13	D3
Store	5		12	$\overline{Q3}$
Polarity	6		11	Q3
D2	7		10	Q2
	8		9	$\overline{Q2}$

4043
QUAD R/S FLIP-FLOP
NOR Logic

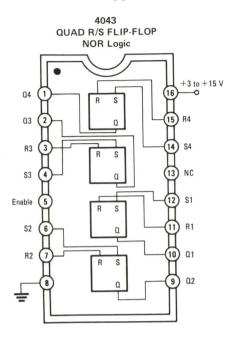

Q4	1	16	+3 to +15 V
Q3	2	15	R4
R3	3	14	S4
S3	4	13	NC
Enable	5	12	S1
S2	6	11	R1
R2	7	10	Q1
	8	9	Q2

4044
QUAD R/S FLIP-FLOP
NAND Logic

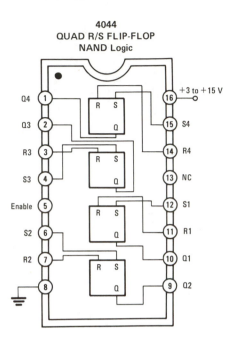

Q4	1	16	+3 to +15 V
Q3	2	15	S4
R3	3	14	R4
S3	4	13	NC
Enable	5	12	S1
S2	6	11	R1
R2	7	10	Q1
	8	9	Q2

4046
PHASE-LOCKED LOOP

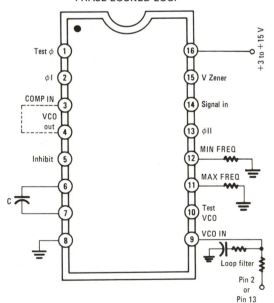

Test ϕ	1	16	+3 to +15 V
ϕI	2	15	V Zener
COMP IN	3	14	Signal in
VCO out	4	13	ϕII
Inhibit	5	12	MIN FREQ
	6	11	MAX FREQ
C	7	10	Test VCO
	8	9	VCO IN

Loop filter

Pin 2
or
Pin 13

4049
HEX INVERTING BUFFER
& TTL DRIVER

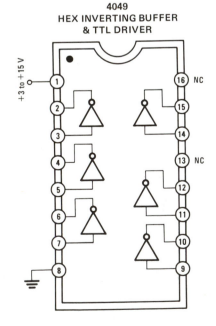

+3 to +15 V	1	16	NC
	2	15	
	3	14	
	4	13	NC
	5	12	
	6	11	
	7	10	
	8	9	

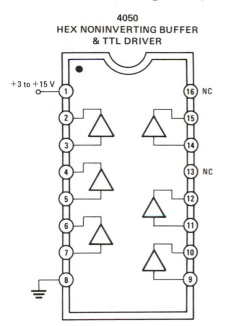

4050
HEX NONINVERTING BUFFER
& TTL DRIVER

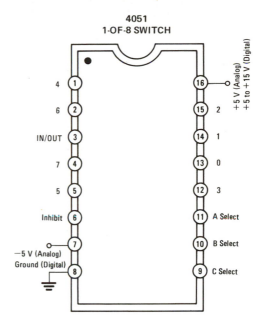

4051
1-OF-8 SWITCH

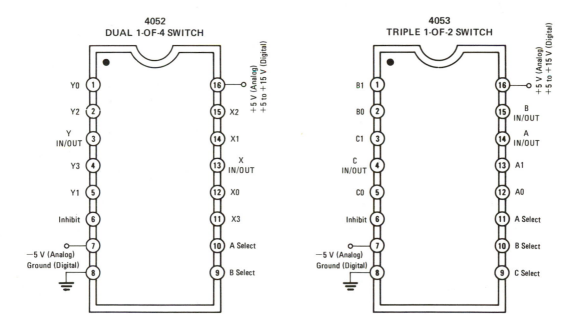

4052
DUAL 1-OF-4 SWITCH

4053
TRIPLE 1-OF-2 SWITCH

4060
14-STAGE (÷16,384) BINARY RIPPLE COUNTER
With Internal Oscillator

Pin			Pin	
12	1		16	+3 to +15 V
13	2		15	10
14	3		14	8
6	4		13	9
5	5		12	Reset
7	6		11	Clock
4	7		10	OSC IN
	8		9	OSC OUT

Outputs (pins 1–7), Outputs (pins 15–13)

4063
4-BIT MAGNITUDE COMPARATOR

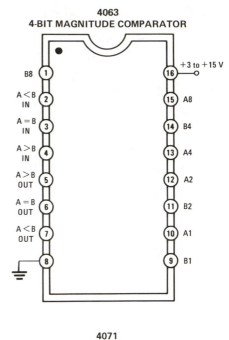

	Pin		Pin	
B8	1		16	+3 to +15 V
A < B IN	2		15	A8
A = B IN	3		14	B4
A > B IN	4		13	A4
A > B OUT	5		12	A2
A = B OUT	6		11	B2
A < B OUT	7		10	A1
	8		9	B1

4067
1-OF-16 ANALOG SWITCH

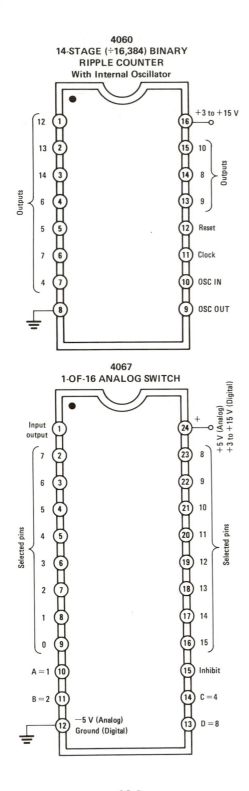

	Pin		Pin	
Input output	1		24	+5 V (Analog) / +3 to +15 V (Digital) +
7	2		23	8
6	3		22	9
5	4		21	10
4	5		20	11
3	6		19	12
2	7		18	13
1	8		17	14
0	9		16	15
A = 1	10		15	Inhibit
B = 2	11		14	C = 4
−5 V (Analog) Ground (Digital)	12		13	D = 8

Selected pins

4071
QUAD 2-INPUT OR GATE

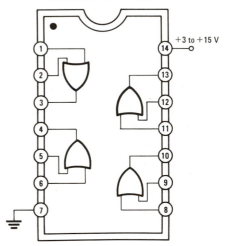

Pin		Pin	
1		14	+3 to +15 V
2		13	
3		12	
4		11	
5		10	
6		9	
7		8	

420

4068
8-INPUT NAND GATE

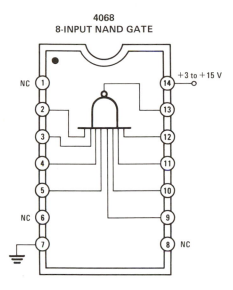

4069
HEX INVERTER

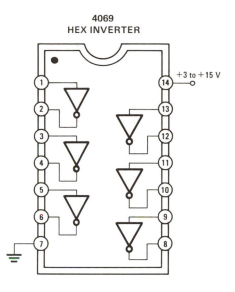

4070
QUAD EXCLUSIVE OR GATE

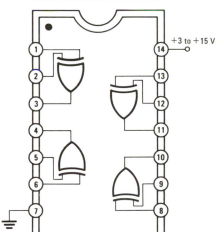

4072
DUAL 4-INPUT OR GATE

4073
TRIPLE 3-INPUT AND GATE

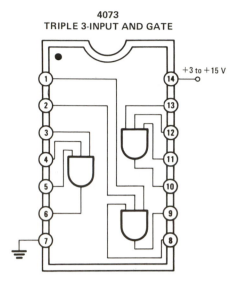

4075
TRIPLE 3-INPUT OR GATE

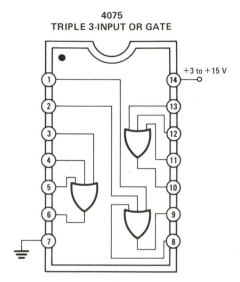

4076
QUAD D REGISTER,
TRI-STATE

Enable
output

Q1
Q2
OUT
Q3
Q4

Clock

+3 to +15 V

Reset

D1
D2
IN
D3
D4

Enable
input

4077
QUAD EXCLUSIVE NOR GATE

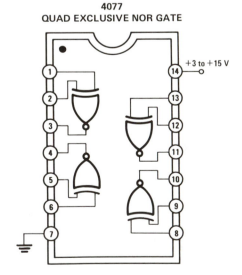

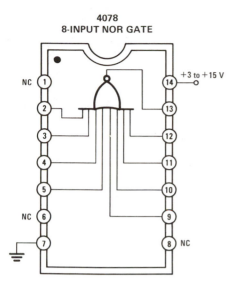

4078
8-INPUT NOR GATE

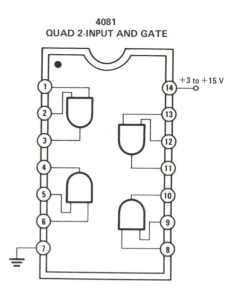

4081
QUAD 2-INPUT AND GATE

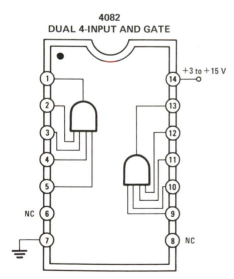

4082
DUAL 4-INPUT AND GATE

4089
BINARY RATE MULTIPLIER

"15" OUT ① — ⑯ +3 to +15 V

Rate { C = 4 ② — ⑮ B = 2 } Rate

D = 8 ③ — ⑭ A = 1

15 Set ④ — ⑬ Clear

$\overline{OUT}$ ⑤ — ⑫ Cascade

OUT ⑥ — ⑪ Inhibit IN

Inhibit OUT ⑦ — ⑩ Strobe

⑧ — ⑨ Clock

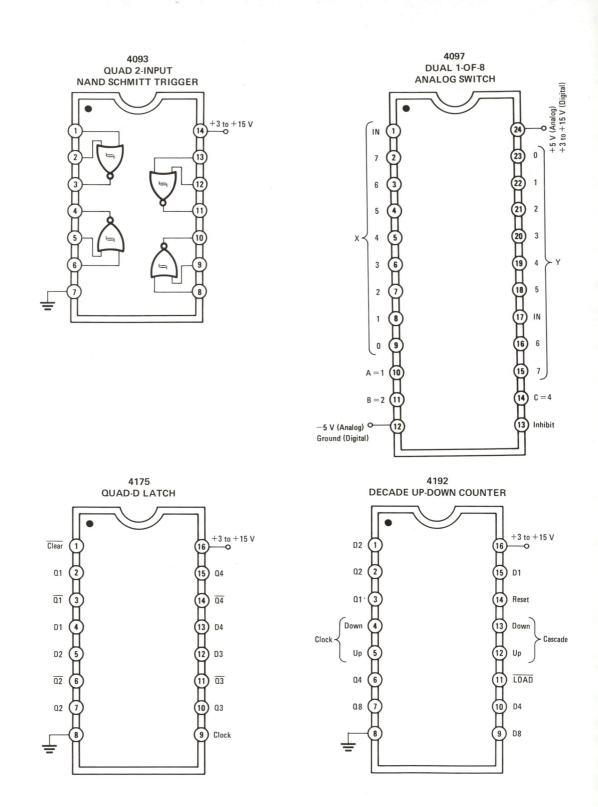

4093
QUAD 2-INPUT
NAND SCHMITT TRIGGER

1
2
3
4
5
6
7
14 +3 to +15 V
13
12
11
10
9
8

4097
DUAL 1-OF-8
ANALOG SWITCH

IN — 1
7 — 2
6 — 3
5 — 4
X { 4 — 5
3 — 6
2 — 7
1 — 8
0 — 9
A = 1 — 10
B = 2 — 11
−5 V (Analog) — 12
Ground (Digital)

24 — +5 V (Analog)
       +3 to +15 V (Digital)
23 — 0
22 — 1
21 — 2
20 — 3
19 — 4 } Y
18 — 5
17 — IN
16 — 6
15 — 7
14 — C = 4
13 — Inhibit

4175
QUAD-D LATCH

$\overline{\text{Clear}}$ — 1
Q1 — 2
$\overline{\text{Q1}}$ — 3
D1 — 4
D2 — 5
$\overline{\text{Q2}}$ — 6
Q2 — 7
8

16 — +3 to +15 V
15 — Q4
14 — $\overline{\text{Q4}}$
13 — D4
12 — D3
11 — $\overline{\text{Q3}}$
10 — Q3
9 — Clock

4192
DECADE UP-DOWN COUNTER

D2 — 1
Q2 — 2
Q1 — 3
Down — 4 } Clock
Up — 5
Q4 — 6
Q8 — 7
8

16 — +3 to +15 V
15 — D1
14 — Reset
13 — Down } Cascade
12 — Up
11 — $\overline{\text{LOAD}}$
10 — D4
9 — D8

424

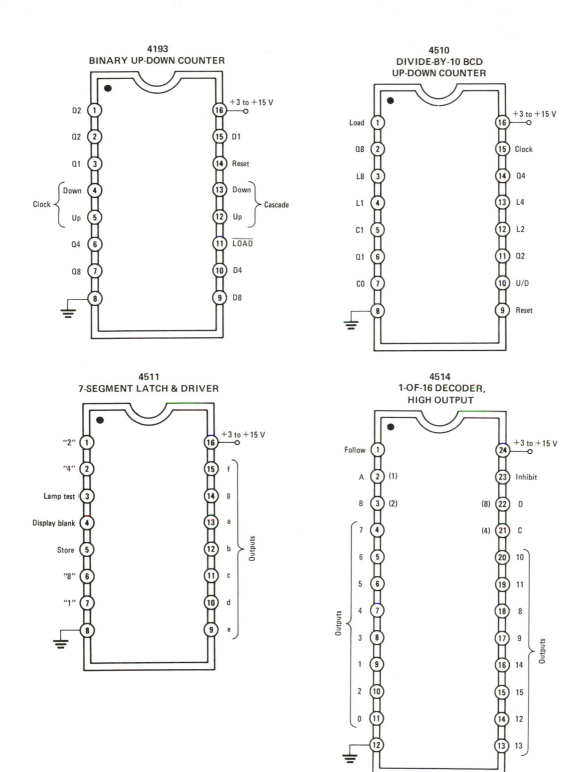

4193
BINARY UP-DOWN COUNTER

D2	1	16	+3 to +15 V
Q2	2	15	D1
Q1	3	14	Reset
Down	4	13	Down
Up	5	12	Up
Q4	6	11	LOAD
Q8	7	10	D4
	8	9	D8

Clock { Down, Up }
Cascade { Down, Up }

4510
DIVIDE-BY-10 BCD
UP-DOWN COUNTER

Load	1	16	+3 to +15 V
Q8	2	15	Clock
L8	3	14	Q4
L1	4	13	L4
C1	5	12	L2
Q1	6	11	Q2
C0	7	10	U/D
	8	9	Reset

4511
7-SEGMENT LATCH & DRIVER

"2"	1	16	+3 to +15 V
"4"	2	15	f
Lamp test	3	14	g
Display blank	4	13	a
Store	5	12	b
"8"	6	11	c
"1"	7	10	d
	8	9	e

Outputs

4514
1-OF-16 DECODER,
HIGH OUTPUT

Follow	1	24	+3 to +15 V
A	2 (1)	23	Inhibit
B	3 (2)	22	(8) D
7	4	21	(4) C
6	5	20	10
5	6	19	11
4	7	18	8
3	8	17	9
1	9	16	14
2	10	15	15
0	11	14	12
	12	13	13

Outputs

Outputs

425

4515
1-OF-16 DECODER,
LOW OUTPUT

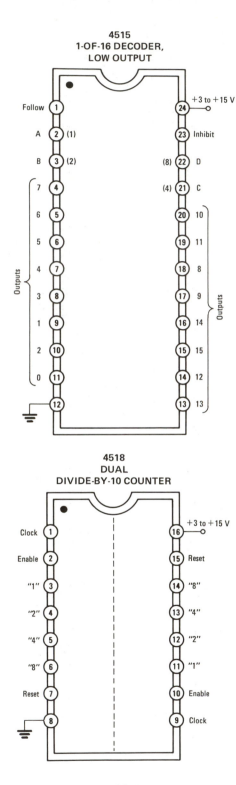

Follow	1	24	+3 to +15 V
A	2 (1)	23	Inhibit
B	3 (2)	(8) 22	D
7	4	(4) 21	C
6	5	20	10
5	6	19	11
4	7	18	8
3	8	17	9
1	9	16	14
2	10	15	15
0	11	14	12
	12	13	13

Outputs (left bracket 4–11)

Outputs (right bracket 17–13)

4516
DIVIDE-BY-16 BINARY
UP-DOWN COUNTER

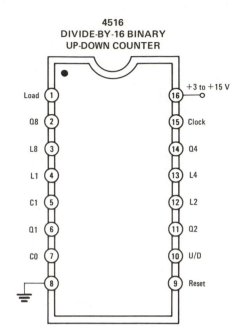

Load	1	16	+3 to +15 V
Q8	2	15	Clock
L8	3	14	Q4
L1	4	13	L4
C1	5	12	L2
Q1	6	11	Q2
C0	7	10	U/D
	8	9	Reset

4518
DUAL
DIVIDE-BY-10 COUNTER

Clock	1	16	+3 to +15 V
Enable	2	15	Reset
"1"	3	14	"8"
"2"	4	13	"4"
"4"	5	12	"2"
"8"	6	11	"1"
Reset	7	10	Enable
	8	9	Clock

4520
DUAL
DIVIDE-BY-16 COUNTER

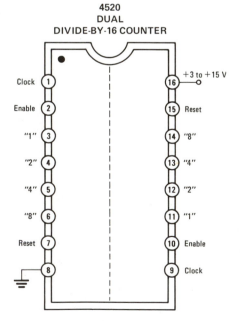

Clock	1	16	+3 to +15 V
Enable	2	15	Reset
"1"	3	14	"8"
"2"	4	13	"4"
"4"	5	12	"2"
"8"	6	11	"1"
Reset	7	10	Enable
	8	9	Clock

426

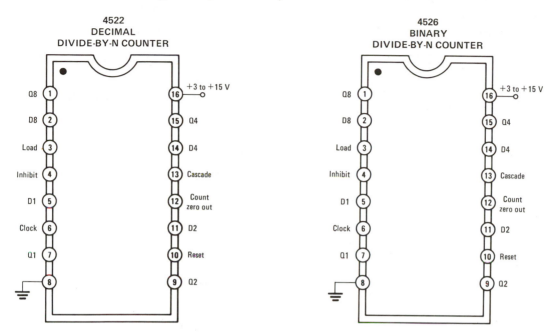

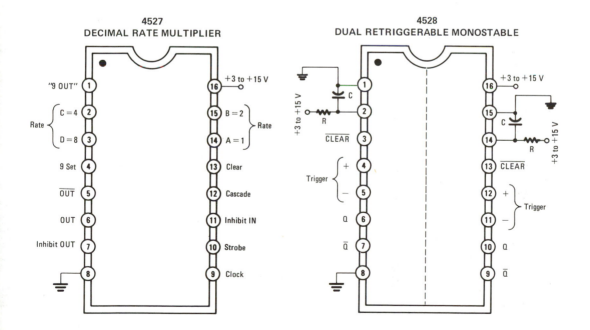

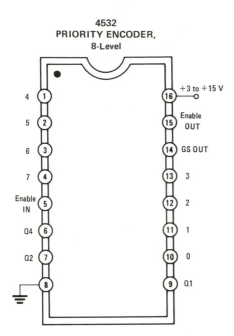

4532
PRIORITY ENCODER,
8-Level

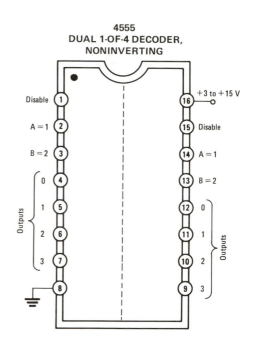

4555
DUAL 1-OF-4 DECODER,
NONINVERTING

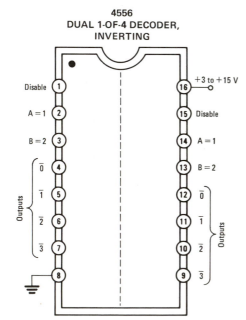

4556
DUAL 1-OF-4 DECODER,
INVERTING

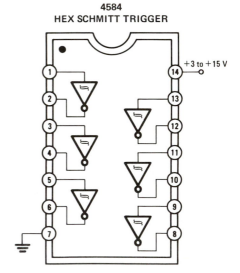

4584
HEX SCHMITT TRIGGER

APPENDIX C: REQUIRED PARTS AND EQUIPMENT FOR THE EXPERIMENTS

Resistors

7	220 Ω	1	10 kΩ	2	100 kΩ
1	330 Ω	2	24 kΩ	1	220 kΩ
1	560 Ω	1	27 kΩ	1	330 kΩ
4	1 kΩ	1	39 kΩ	1	390 kΩ
1	4.7 kΩ	2	56 kΩ	1	1 MΩ

All resistors 1/4 watt, 5%

Capacitors

1	0.001 μF	3	0.1 μF	1	1 μF
1	0.01 μF	1	0.33 μF	1	10 μF
2	0.082 μF				

CMOS Devices

1	4001 quad 2-input NOR gate
1	4017 decoded 1-of-N counter
1	4018 five-stage Johnson counter

CMOS Devices (*Continued*)

1   4046 phase-locked loop

1   4047 multivibrator

TTL Devices

1   555 timer

1   7400 quad 2-input NAND gate

1   7402 quad 2-input NOR gate

1   7404 hex inverter

1   7408 quad 2-input AND gate

1   7432 quad 2-input OR gate

1   7442 BCD-to-decimal decoder

1   7447 BCD-to-7-segment decoder/driver

1   7474 dual *D*-type flip-flop with preset and clear

1   7475 4-bit latch

1   7476 dual *J–K* flip-flop with preset and clear

1   7483 4-bit parallel adder

1   7486 quad XOR gate

1   7489 16 $\times$ 4 bit RAM

1   7490 decade counter

1   7493 4-bit binary (divide-by-16) counter

2   74126 quad noninverting three-state buffer

1   74150 16-line multiplexer

1   74154 16-line demultiplexer

1   74164 serial in–parallel out shift register

1   74192 programmable up–down counter

Miscellaneous

8   TTL logic switches

2   TTL logic pulsers

1   CMOS logic pulser

8   TTL LED monitors

Miscellaneous (*Continued*)

1   square-wave clock

1   seven-segment LED display with decoder/driver

1   LED (red, green, orange, or yellow lens)

1   MAN-7 common-anode seven-segment LED (or equivalent)

1   1N914 silicon diode

1   two-channel oscilloscope

1   function generator

1   frequency counter

1   dc voltmeter

1   0 to 15 volt dc power supply

1   solderless breadboarding socket

    wire

INDEX